CONCEPTS IN ANATOMY

CONCEPTS IN ANATOMY

by

HERBERT H. SREBNIK, Ph.D.
University of California, Berkeley

Illustrations by
Genevieve M. Wilson, CMI

KLUWER ACADEMIC PUBLISHERS
Boston / Dordrecht / London

Distributors for North, Central and South America:
Kluwer Academic Publishers
101 Philip Drive
Assinippi Park
Norwell, Massachusetts 02061 USA
Telephone (781) 871-6600
Fax (781) 681-9045
E-Mail <kluwer@wkap.com>

Distributors for all other countries:
Kluwer Academic Publishers Group
Distribution Centre
Post Office Box 322
3300 AH Dordrecht, THE NETHERLANDS
Telephone 31 78 6392 392
Fax 31 78 6546 474
E-Mail <services@wkap.nl>

Electronic Services <http://www.wkap.nl>

Library of Congress Cataloging-in-Publication Data

Srebnik, Herbert H., 1923-
Concepts in anatomy / by Herbert H. Srebnik; illustrations by Genevieve M. Wilson
p.cm.
Includes bibliographicl references and index.
ISBN 0-7923-7539-4 (alk. paper)
1. Human anatomy. I. Title

QM23.2 .S695 2001
611—dc21

2001038760

Printed on acid-free paper.
Printed in the United States of America

The Publisher offers discounts on this book for course use and bulk purchases. For further information, send email to <joanne.tracy@wkap.com>.

DEDICATION

This book is dedicated to the memory of

Virginia Rose Harris (1908–1999)

who never failed me

CONTENTS

PREFACE

Concepts in Anatomy is not a textbook of anatomy; there are already many excellent anatomy texts on the market, each with its own unique perspective and point of view. *Concepts* is a collection of lecture notes compiled, edited, and distilled over many years of teaching human developmental and gross anatomy to undergraduate, graduate, and medical students. It aims to speak to you, the audience, directly, clearly, and succinctly.

Concepts is less than a textbook, but also more than a synopsis. It is selective rather than encyclopedic in scope, and conceptual rather than particular in its approach. It stresses general principles, so as to minimize rote memory learning, and it provides order and direction to the study of gross anatomy. *Concepts* attempts to help you make sense of subject matter that is inherently complicated or confusing—and to make you aware that knowledge and application of developmental anatomy frequently renders such difficult material comprehensible. Needless to say, I have drawn attention at every opportunity to the close relationship of form and function.

The overall arrangement of *Concepts* and its contents reflect the sequence in which parts of the human body were studied in the gross anatomy laboratories at the University of California at Berkeley. Individual chapters are fairly independent of each other, however, and the order of their arrangement is open to revision. An elemental knowledge of human structure is assumed, as is knowledge of when, how, and where to obtain supplementary information.

Some chapters target areas of systemic anatomy, *e.g.*, joint typology, or the membranes of the body; others deal with important aspects of functional integration, including patterns of collateral circulation, fields of lymphatic drainage, and the relevance of neuromuscular units. Several chapters are devoted to the developmental history of organ system in order to explain shape, location, and final disposition of adult structure. Many focus on topics that gave me trouble when I was a student—among them, the myofascial anatomy of the perineum. Still others suggest strategies to help reduce the tedium of onerous detail; the assertion made by the great French artist, Henri Matisse, in *Notes of a Painter* (1908), that "All that is not essential to a painting is detrimental to it" should inform also the teaching of descriptive anatomy.

Anatomy, to be sure, is the essential foundation of clinical practice, but it is much more than that. First and foremost, anatomy is a biological science. There is order and logic to the organization of the human body and the arrangement of its parts. And, as all science, anatomy offers challenge and discovery. It exercises and stretches the mind and fosters scholarly inquiry. It will enrich your lives; for, as an aesthetic expression, anatomy is a powerful and beautiful subject that introduces you to the magic and majesty of the human body.

In the end, *Concepts in Anatomy* is an innovative, imaginative, and informative study aid. At a time when instruction in human anatomy has become an endangered species, this book may offer compensation for those of you majoring in biology or pursuing careers in the health sciences. I also view *Concepts* as a refresher course for interns and residents training for board certification, and for clinicians wishing to review the fundamentals of human gross anatomy. Physicians, health professionals in allied fields, including practitioners of alternative medicine and medical illustrators, may find this book a rich source of information and, possibly, even inspiration.

Herbert H. Srebnik, Ph.D.
Berkeley, California

ACKNOWLEDGMENTS

This venture probably would have come never to fruition without the encouragement of former and present students. I am greatly indebted to them for planting the seed. Genevieve Wilson, by now an accomplished medical illustrator, insisted early in her training on doing the art work. Her unique talents are evident everywhere in this book. Some of her drawings are based on diagrams by the late Ian W. Monie, M.D., Professor of Anatomy, University of California at Berkeley and San Francisco. Indeed, I wish to acknowledge the impact of my own teachers, Professor Monie included, whose ideas and insights have shaped my career as human anatomist and form the staple of my lectures and demonstrations. Likewise, I should have accomplished little in my professional life without the wisdom and efforts of untold numbers of authors whose textbooks, monographs, atlases, and syllabi I have consulted in the course of many years of teaching. Several illustrations from those sources have been adapted for use in this book.

I owe a special debt to Theodora Crawford, who reviewed and readied the manuscript for publication with admirable skill and perspicacity; and to Aubrey McClellan who prepared the index .

Among the many individuals who took an active interest in improving the quality of the text, Larry Elson and Becky Hainz-Baxter deserve to be singled out for their constructive comments and suggestions.

It is my pleasure to commend Joanne Tracy, Biosciences editor at Kluwer Academic Publishers, and her staff for their commitment to the project and its realization.

Last, but not least, I thank my wife, Marilyn, and our children, Laura and David, for providing much needed comfort, support, and tender loving care during the many months of writing.

Chapter 1
INTRODUCTION

The word ***anatomy*** comes from the Greek (*ana-tome*, to cut up). It refers to the form and arrangements of parts into which the body may be rendered by anatomizing or dissection (from the Latin *dis-secare*, to cut apart). There is no such word as *di-section* .

According to the *Anatomica Ricardi*, a description of human anatomy published in Salerno during the 12th century: "A knowledge of anatomy is necessary to physicians, in order that they may understand how the human body is constructed to perform different movements and operations." In the same vein, Andreas Vesalius of Brussels (1514–1564), the father of modern human anatomy, wrote: "Anatomy should rightly be regarded as the firm foundation of the whole art of medicine and its essential preliminary."

Vesalius began his anatomical studies in Louvain and Paris. To further his education, he set off for the University of Padua, then recognized as one of the great centers of learning. At age 23, after obtaining a medical degree, Vesalius was appointed Professor of Surgery and Anatomy at that university. Convinced that the teachings of earlier anatomists were based, almost entirely, on observations of apes and other animals rather than humans, Vesalius embarked on a systematic dissection of human cadavers. The result, a defining moment in the history of human anatomy, was the publication of *De Humani Corporis Fabrica* (the Structure of the Human Body) in 1543, a magnificent account of his work consisting of over seven hundred pages of text and illustrations.

Subdivisions of Anatomy

Gross anatomy relies heavily on dissection for identification and examination of parts, and **regional** (topographic) **anatomy** is the study of parts of the body, one after another, rather than of the systems of the body. Above all, regional anatomy is the study of spatial relationships; *i.e.*, the definition of an organ's position or a structure's passage with respect to its surroundings. To define position or passage, ask yourself: What lies immediately above organ X, and what lies immediately below it? Or, what passes in front of structure Y, and what passes behind it? Or, what borders space Z on the right, and what borders it on the left?

Systemic (systematic) **anatomy** is the study of the body's individual systems, *e.g.*, skeletal or cardiovascular, rather than of its regions, such as head or neck. Systemic anatomy lets you not lose sight that the body is more than the sum of its parts.

Cross-sectional anatomy is a regional approach that reveals the form and arrangement of structures in slices of the body isolated from the rest and undisturbed by dissection.

Developmental anatomy deals with phases of prenatal life (embryology), but also with postnatal development; for growth and differentiation do not cease at birth. Beyond that, a working knowledge of developmental anatomy renders an organ's ultimate shape, structure, and location in the adult body comprehensible, thereby substantially reducing mind-numbing memorization.

Teratology is the branch of developmental biology concerned with the causes, incidence, and other particulars of birth defects and congenital malformations.

Neuroanatomy is all about the development, structure, and function of the central nervous system; *i.e.*, the brain and spinal cord. The brain is perhaps the most striking example of an organ whose ultimate shape and structure are best explained through knowledge of its developmental history. We here, however, will touch upon it only tangentially.

Microscopic anatomy delves into the structure of the body's fabrics not visible with the naked eye. Histology is the study of the tissues of the body, cytology is the examination of cell structure, and electronmicroscopy attempts the elucidation of subcellular detail. Microanatomy, too, is largely beyond the scope of this discourse.

Radiologic anatomy brings to light parts of the body normally hidden from view. Conventional methods utilize penetrating X-ray beams which, depending on the density and thickness of the object through which they pass, expose either partially or completely a photographic film placed behind the object. More recent technology includes CT (computer-assisted tomography) and MRI (magnetic resonance imaging) scans.

Surface anatomy relates the body's external contours or features to structures beneath the skin. It also refers to the art of projecting internal organs—their outlines, shapes, or relationships—onto the surface of the body with the aid of a marking pen. Its methodology includes direct observation, palpation (sense of touch), and percussion. Surface anatomy should appeal to anybody interested in *living anatomy*, including painters and sculptors of the human figure.

Clinical anatomy emphazizes the practical application of the subject. It is part of every health professional's training and repertory.

Conceptual anatomy is a method of creative and critical thinking that leads to the formulation of general principles. It is a process by which seemingly unrelated observations are tied together into a conceptual framework. Its application to gross anatomy simplifies retention of information; for, once you have assimilated the fundamentals of a basic plan, only exceptions to it need be remembered. To illustrate: All muscles moving the tongue are supplied by the same cranial nerve (N. XII*), except the palatoglossus muscle* (N. X–XI); or, all long bones of the body are preformed in cartilage, *except the clavicle*, which develops in membrane; or, all non-voluntary sphincters of the body (*e.g.*, of bladder and rectum) relax under the influence of parasympathetic nerves, *except the pupillary sphincter,* which contracts.

Rules, Conventions, and Terminology

To negotiate descriptive anatomy successfully, and to communicate with others effectively, you must become thoroughly and rapidly familiar with its language and frames of reference.*

In the **anatomical position** (Fig. 1.1), the body is erect; eyes are level, arms are suspended and hands kept open with palms facing forward; legs are extended with toes pointing straight ahead. No matter what their actual location in space may be, parts of the body are always described in this anatomically correct position; thus, even when you stand on your head, the neck is below (inferior to) the head, not above it. Likewise, when you are flat on your face, the chest is in front of (anterior to) the backbone, not below it.

The **planes** of the body (Fig. 1.2) are imaginary but useful means to describe orientation of the body and its parts with respect to horizontal and vertical axes. The **median** (midsagittal, or just sagittal) **plane** bisects the body longitudinally into right and left halves; **paramedian** (or parasagittal) **planes** , then, are planes parallel to the median plane. **Frontal** (coronal) **planes** pass at right angles to the median plane, thus slice the body front to back. **Horizontal** (transverse) **planes** cut across the long axis of the body or its parts; they are the basis of cross-sectional anatomy.

Terms that define direction and location (Fig. 1.3) usually come in pairs, the partners of which express opposites:

Anterior and *Posterior*—closer to the front vs. closer to the back

Superior and *Inferior*—closer to the top vs. closer to the ground

Note that these terms are descriptive only when the body is in the anatomical position! You may substitute, however, *ventral-dorsal* for *anterior-posterior*, or use *cranial-caudal* instead of *superior-inferior. Cranial, cephalic,* and *rostral* all indicate head end; *caudal* refers to tail end. The terms *ventral-dorsal* and *cranial-caudal* are obligatory in embryology and comparative anatomy; for, in early human embryonic stages and four-footed animals, the term anterior is equivalent to head end and posterior to tail end. Other descriptive terms are:

Medial and *Lateral*—closer to the median plane vs. farther from it

Deep and *Superficial*—closer to the body's surface vs. farther from it

Internal and *External*—closer to the center of a cavity vs. farther from it

Proximal and *Distal*—nearer to vs. farther from a structure's root. These terms usually express relative distance from the source, *e.g.*, the shoulder is the proximal part of the upper limb, the hand its distal part.

Other pairs of terms describe actions and modalities of movement of one part of the body upon another (Fig. 1.4):

Flexion and *Extension*—to bend *vs.* to straighten

Abduction and *Adduction*—away from *vs.* towards the median plane

Lateral and *Medial Rotation*—turning away from vs turning towards the median plane. These terms are commonly used to describe limb rotation about its long axis. *External-Internal* may replace *Lateral-Medial.*

Circumduction—a circular movement combining all of the modalities above, exemplified by having the extended arm trace out the base of a cone with the shoulder joint as its apex

Supination and *Pronation*—face up vs. face down. A pair of terms indicating lateral and medial rotation, respectively, of the radius on the ulna—causing the palm of the hand to face front or back in the Anatomical Position.

The terms *Origin* and *Insertion* describe muscle attachments to bone. The *origin* of a muscle is the end closer to the median plane; the *insertion* is the end farther away from it. Alternatively, origin of a muscle may be defined as the end that is attached to the fixed bone of a pair—*i.e.*, the end that remains stationary as the muscle shortens; and insertion is the part of a muscle that moves the other bone of the pair through distance. Based on this definition, a muscle that moves one bone of a pair at one time and the other bone of the pair at another time will reverse also its origin and insertion (Fig. 1.5).

* The terminology used in this book conforms largely with the official terminology adopted by the International Federation of Associations of Anatomists (IFAA) in July, 1998.

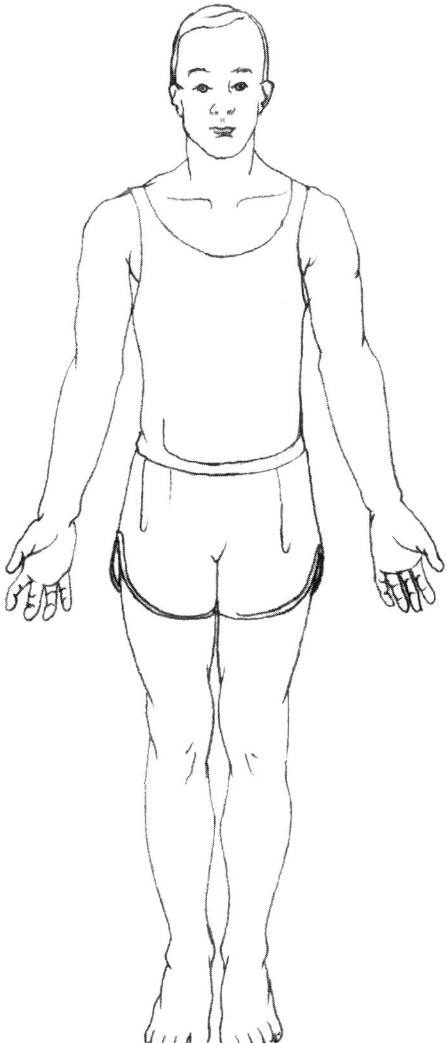
1.1 Anatomical position

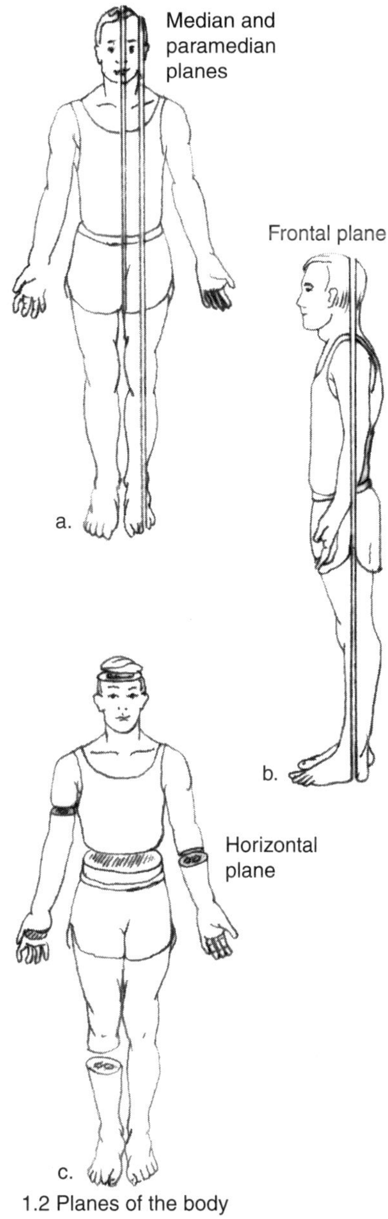

1.2 Planes of the body

Figure 1.1: Anatomical position
Figure 1.2: Planes of the body

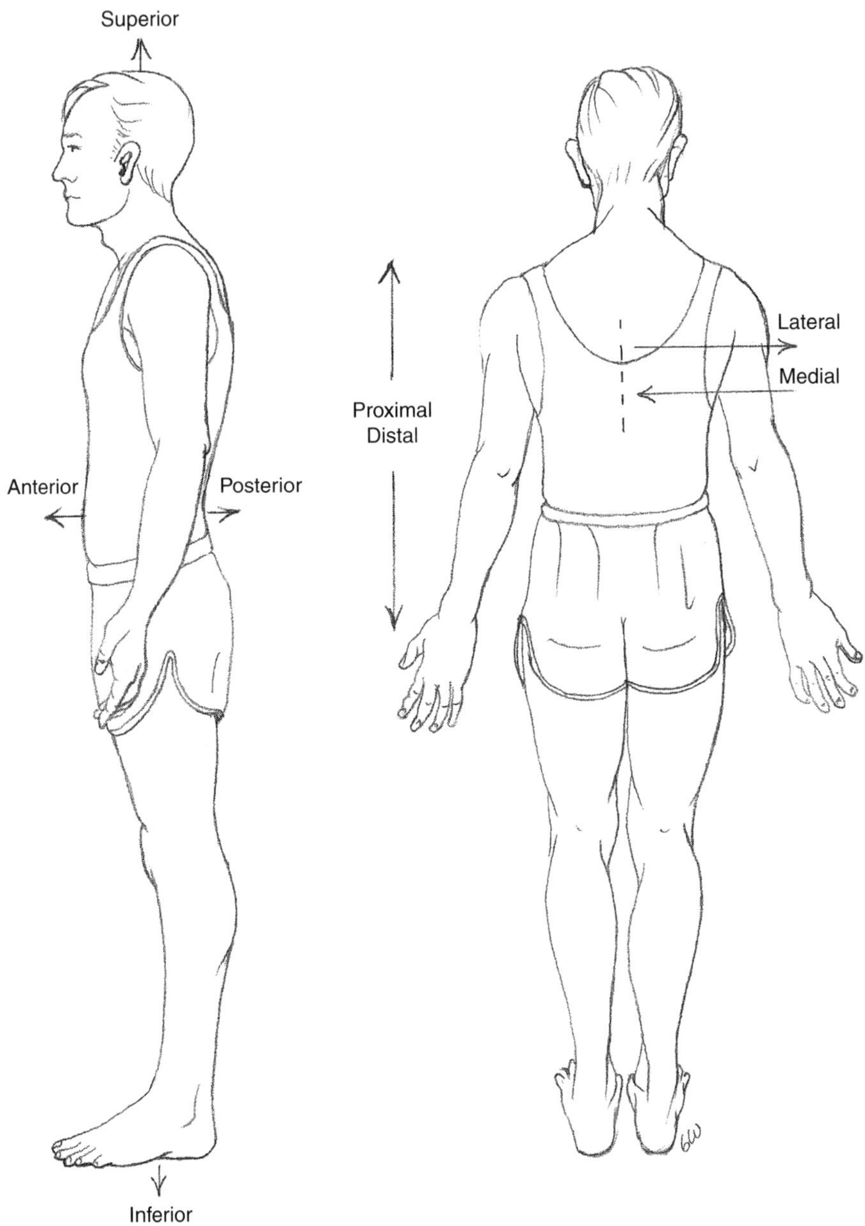

Figure 1.3: Terms defining direction and location

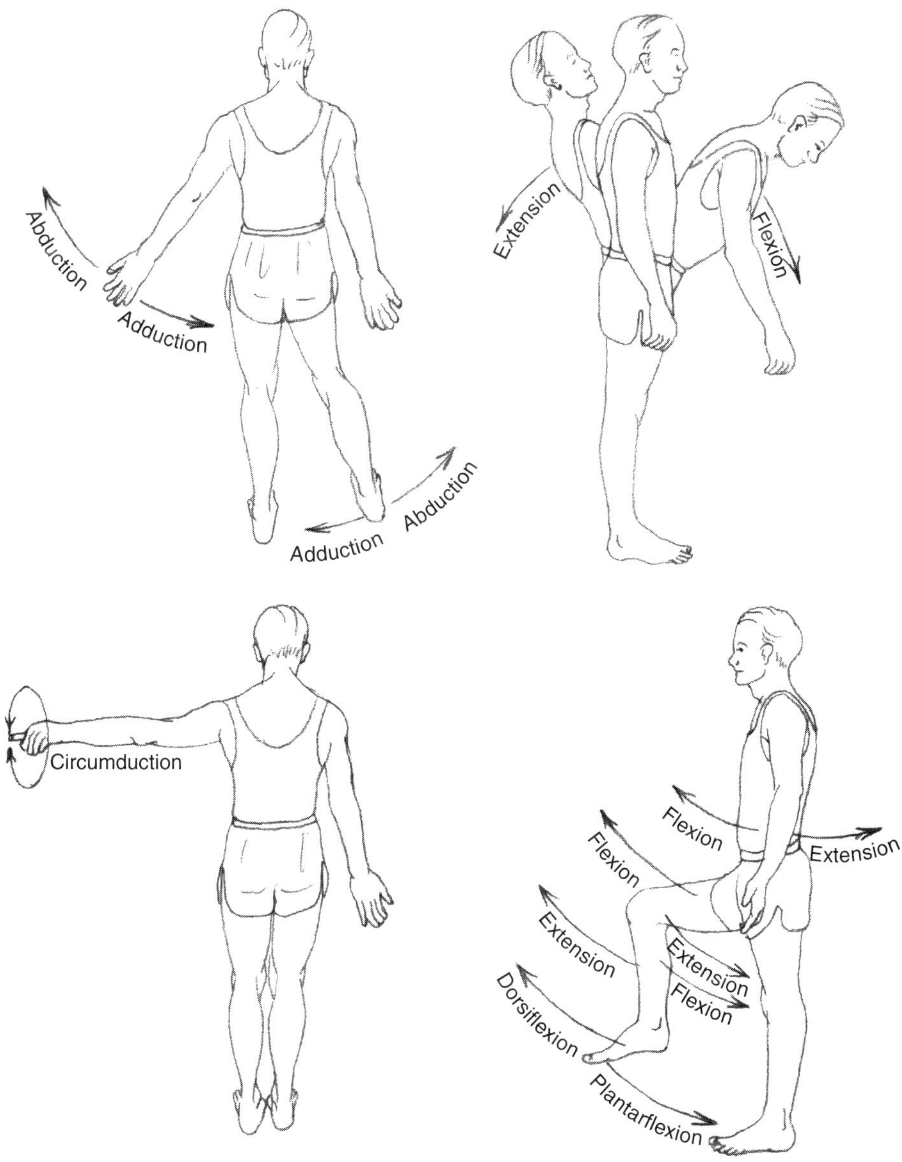

Figure 1.4a: Terms expressing action and motion

1.4b Terms expressing action and motion

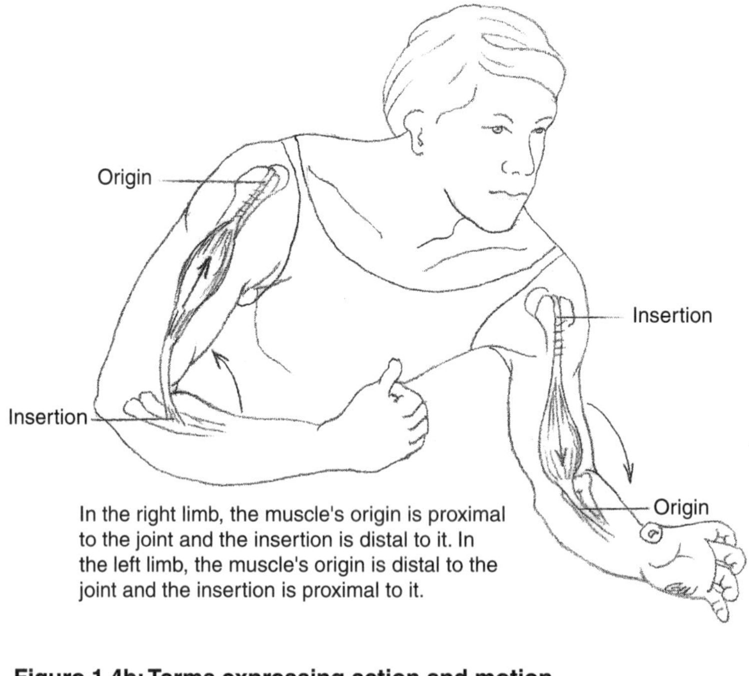

Figure 1.4b: Terms expressing action and motion
Figure 1.5: Reversal of muscle origin and insertion

Chapter 2
ON FORM AND FUNCTION

A great deal has been written on this subject, and you are probably familiar with the expression "Form follows Function"—meaning, structure is designed for the use to which it is put. Leonardo da Vinci (1452–1519), quite possibly the most creative genius of his day, referring to the relationship between muscle size and function, concluded: "In all the parts where man has to work with greater effort nature has made the muscles and tendons of greater thickness and breadth" (*Dell' Anatomica Fogli A*, Royal Library, Windsor). The musculoskeletal anatomy of the limbs is particularly well suited to illustrate and illuminate the anatomical basis of function, for both upper and lower extremities are structurally adapted for maximal effectiveness: the former to be used as a prehensile organ, and the latter to provide the means of support for upright posture and locomotion. Crucial to an understanding of their functional capacities is a knowledge of their bony infrastructure.

The Appendicular Skeleton

The appendicular skeleton is composed of paired upper and lower limb girdles and the bones of the upper and lower extremities. The **shoulder** or **pectoral girdle**, attaching the upper limb to the axial skeleton, consists of clavicle and scapula. The lower, **pelvic girdle** is the hip bone (os coxae), formerly known as the innominate bone. The two hip bones attach behind to the sacrum, *i.e.*, the lower end of the axial skeleton, and they join in front to form the pubic symphysis. Early in life, each os coxae consists of three separate bones: pubis, ilium, and ischium; they fuse to become a single unit at or about age 15. Upper limb bones are those of the arm, forearm, wrist, and hand; the bones of the lower limb include those of the thigh, leg, ankle, and foot.

Comparison of Upper and Lower Limbs

The most important function of the upper limb is to place the hand in a position where it can be used to grasp, grip, explore, or manipulate—and the anatomy of the limb is designed to make that possible. Thus, the upper limb is only loosely attached to the axial skeleton, its single bony connection being the joint between clavicle and sternum (Fig. 2.1). Further, the clavicle—phylogenetically a bone of relatively recent development—tends to push the shoulder joint away from the trunk, thereby giving it more space to move freely. The shoulder joint, between scapula and humerus, itself is highly mobile; for the shallow socket of the scapula, the glenoid cavity, does not fully accomodate the large surface area of the humeral head. These particulars enable the upper limbs to act through a wide range of motions.

In contrast, the lower limbs are specialized for weight-bearing and locomotor functions; and, consistent with those activities, their linkage to the rest of the body is much more substantial than that of the upper limbs. It is achieved first through firm bonding of pelvic girdle and sacrum, creating virtually immovable sacroiliac joints (Fig. 2.2). Next, the hip joint is fitted for optimal congruence between the deep, cup-shaped hip-bone socket, the acetabulum, and the near-spherical head of the femur. Numerous muscles and tendons cross the joint and help stabilize it. Add to this a densely fibrous joint capsule, reinforced by strong ligaments, and you will begin to understand how the lower limb is able to support the body's weight and move it over distance. Be aware, however, that strength at the hip is at the expense of mobility: perhaps only professional ballet dancers and gymnasts, schooled in stretching exercises, are able to overcome the severe limitation imposed by hip joint anatomy on limb extension and rotation.

In general, then, stability and mobility are inversely related. At the shoulder, the former is sacrificed for the sake of the latter—which may explain why the humerus is more frequently dislocated than the femur. Here, restraining forces other than fully engaged joint surfaces or capsular ligaments are required to prevent separation of the pectoral girdle from the axial skeleton and dislocation of the shoulder joint. The essential support here is provided almost exclusively by the many muscles that connect axial skeleton and shoulder girdle or stretch between scapula and humerus. Other features of the appendicular skeleton dispose the upper extremity towards mobility and the lower extremity towards stability. For example, the structural adaptations for pronation and supination exist only in the upper limb, where rotation of the radius on the ulna adds flexibility to wrist and hand. In the lower limb, where rigidity rather than flexibility is desired, the bones of the leg, tibia and fibula, are tightly bound and lack the capacity to move on each other. Also, compare the small and nimble bones at the wrist with the bulky and impacted bones that form the joints of the foot (Fig. 2.3); and contrast the long, slender fingers of the hand with the stubby toes. Last but not least, only the thumb is opposable.

The Bony Pelvis

You are aware, no doubt, that the size, shape, and dimensions of the bony pelvis differ in the two sexes. The ilia are less steep and the iliac fossae much flatter in the female, the cavity is roomier and shorter than that of the male.

The pelvic inlet is kidney-shaped rather than heart-shaped, and all its diameters, as well as those of the pelvic outlet, are longer in the female. This sexual dimorphism reflects the child-bearing function of the female. The increased dimensions provide adequate clearance for passage of the fetus through the birth canal, ensure relatively trouble-free delivery, and reduce the risk of serious injury to mother and child. It helps, of course, that the cranial bones of the fetal skull are unfused at birth and can squeeze past each other. This, too, tells you a great deal about the close relationship of form and function.

Joint Structure and Function

Joints are centers of motion, and maximum utility of the prehensile upper limb, in particular, is obtained only when there is full range of motion at the shoulder, elbow, and wrist, and when the joints of the hand perform as required. Much, of course, depends on joint structure, as I pointed out previously: both size and shape of contact areas, as well as bone-to-bone congruence, determine joint mobility and stability.

Equally important in matters of joint function are factors such as the planar orientation of joint surfaces, or their angles of repose, for they provide information about axes of rotation and the types of movement that are possible or permissible. This is true for joints of both appendicular and axial skeletons. Consider, for instance, the vertebral arch joints of the vertebral column: the articular processes uniting thoracic vertebrae are aligned horizontally in the body, an arrangement that favors rotation of the trunk in the horizontal plane. In contrast, lumbar vertebral arch joints are lined up vertically in the parasagittal plane; accordingly, flexion, extension, even side-bending of the lower trunk are not compromised. Rotation of the lumbar spine, however, is severely limited by this orientation and, when forcibly attempted, may damage the joint.

Thus, when seen from a functional perspective, healthy joints are effective facilitators of movement; and, conversely, structural damage to joints invariably produces range-of-motion deficits.

Other Morphological Correlates of Function

While the focus of the foregoing comments has been on joints as units of function, it is important to realize that individual bones also are functional systems. Do not be misled by the dry, bleached appearance of preserved bones; in their living state, they look and feel differently.

Bones come in all sizes, shapes and forms, each appropriate to need. A major property of bones is their strength, rigidity, and resiliency, and long bones, in particular, serve as levers arms whose movement is controlled by muscles that attach to them.

Evidence for the weight-bearing functions of the lower extremities, for example, can be found in the internal architecture of the upper end of the femur. Here, a complicated framework of spongy bone exists in which trabeculae are arranged in structural patterns so organized as to resist compression forces or other mechanical stresses. In the upper limb, where supination and pronation are major determinants of function, the power of the radius to rotate the forearm is greatly increased by the bone's sinuous contour. The tendons of biceps brachii and pronator teres muscles are inserted into the summits of the radius' curvatures, and their whipping actions as supinator or pronator, respectively, are thus greatly enhanced. Not surprisingly, in gibbons, whose mode of travel includes brachiating through trees, the radius is bowed in the extreme. In short, pay attention to the physical dimensions and shapes of bones—also their markings, *e.g.*, lines, crests, or tubercles, for they serve to attach muscles. You can learn much about power, speed, and efficiency of a muscle crossing a joint by taking into account the length of the lever arm measured from the joint axis to the point of the muscle's insertion.

Before leaving the subject of form and function, let me point out that muscles do not always perform to best advantage when they cross relatively unstable, freely movable joints. Thus, biceps brachii, commonly considered to be the main flexor of the forearm, is hardly that. It crosses the shoulder joint, arguably the most mobile joint in the body, as well as the elbow joint before inserting on the radius—a bone better suited to rotation than flexion. Biceps brachii, therefore, first and foremost is a supinator of the forearm. The principal flexor of the forearm is brachialis, a muscle that arises from the shaft of the humerus and inserts high up on the ulna. The one joint that brachialis crosses, the humero-ulnar joint, is a true hinge—an example of a lever first class. Brachialis, therefore, holding the mechanical advantage over biceps brachii, is a fast, efficient, and effective flexor of the elbow joint.

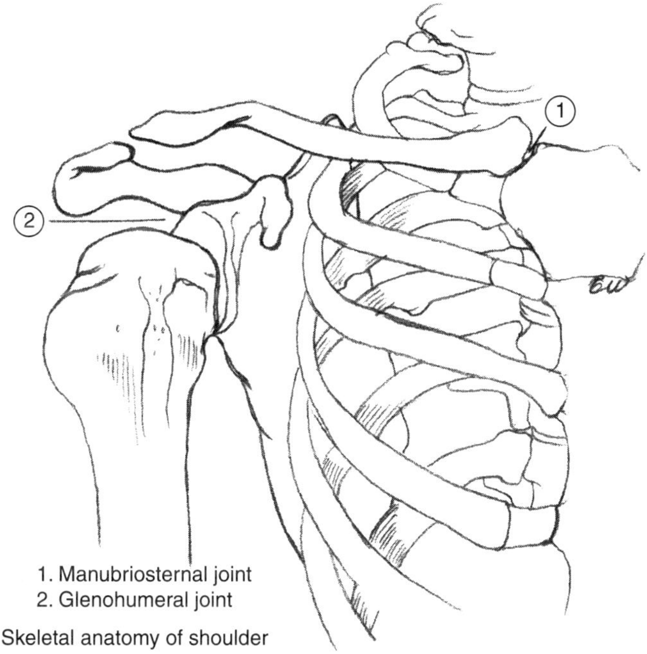

2.1 Skeletal anatomy of shoulder

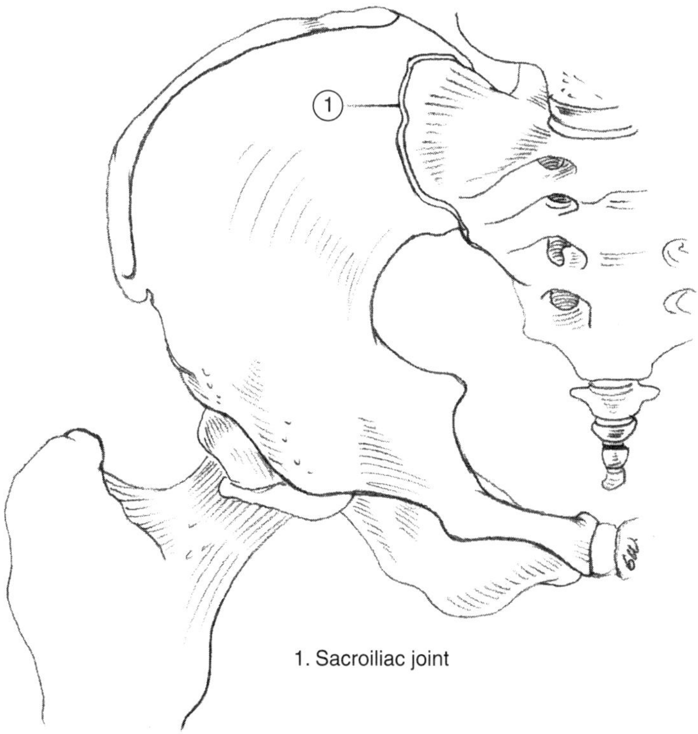

2.2 Skeletal anatomy of hip

Figure 2.1: Skeletal anatomy of shoulder
Figure 2.2: Skeletal anatomy of hip

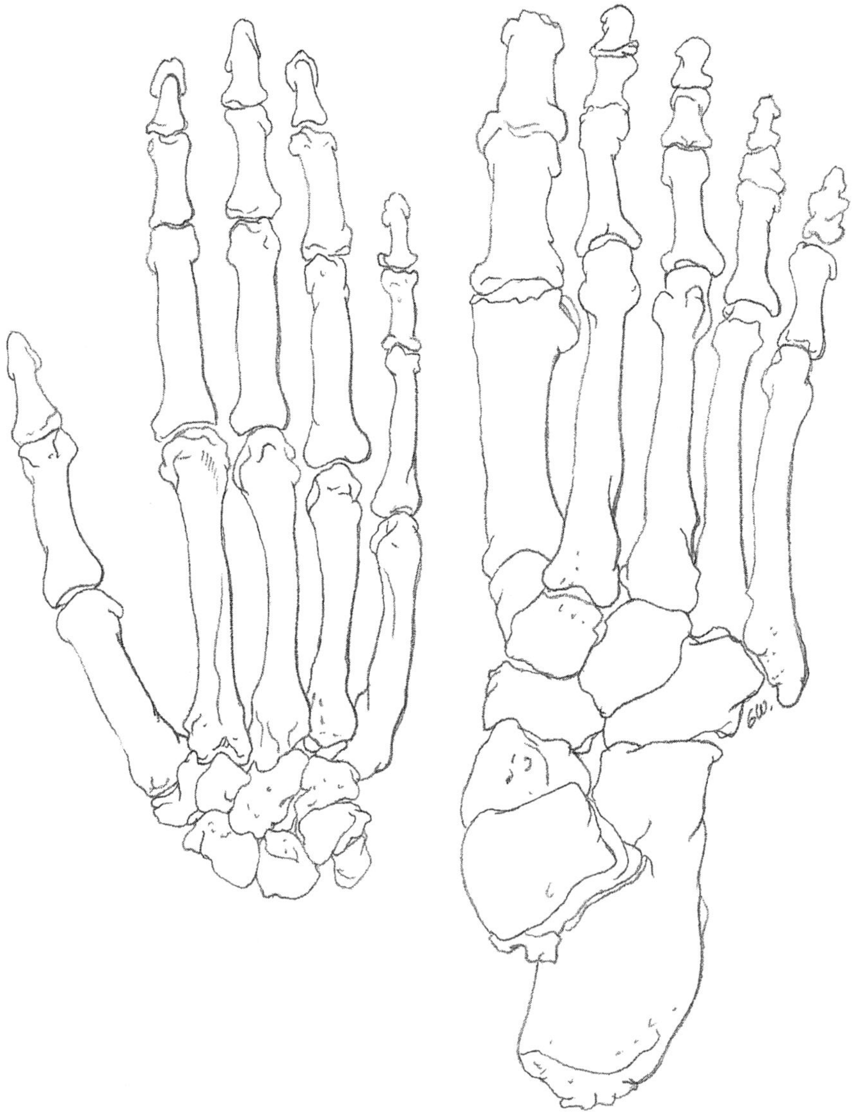

2.3 Skeletal anatomy of hand and foot

Figure 2.3: Skeletal anatomy of the hand and foot

Chapter 3
EMBRYOGENESIS

In the final analysis, human form and function are the end-product of successful embryonic development, which begins with **syngamy** (Gr., *syn* = together; *gamos* = marriage)—the physical union of male and female germ cells or **gametes**. Stringent parameters of time, place, and manner exist for the initiation of pregnancy, including a narrowly scripted program of hormonal support for the attainment of a uterine receptive state following conception.

Syngamy

Syngamy requires a fertilizable egg (ovum) and a fertilizing spermatozoon. The former is supplied when the woman ovulates, typically 14 days before onset of her next menstrual period. Ovum-sperm encounter normally occurs in the ampulla of the uterine (Fallopian) tube; and, once the fertilizing spermatozoon arrives inside the ovoplasm, syngamy is complete. The product of conception is the **zygote**, a single cell—though, the term is frequently applied to all early embryonic stages.

Preimplantation Stages of Pregnancy

Fertilization is a major event in procreation, for it restores the genotype, determines the genetic sex of the embryo, and triggers cell division. The first mitotic division (or cleavage) of the zygote, leading to the two-cell stage, follows fertilization in a matter of hours—it marks the beginning of a phase in development known as **segmentation** (Fig. 3.1–3.2). The next few cell divisions take place while the zygote is being transported, conveyor-belt fashion, through the uterine tube to the uterus. The ball of cells produced in transit is the **morula**. As it enters the uterus, some 96 hours after conception, the morula is still no larger than the unfertilized egg; its diameter is approximately 120 microns (0.120 mm)—about the size of a small pin-head. Subsequent segmentation, however, leads to enlargement of the morula and the development of a central cavity. The morula is renamed **blastocyst**, and the fluid-filled central cavity becomes the **segmentation cavity** or **blastocele**. Cells designated **trophoblast** form its periphery. Projecting into the segmentation cavity from one pole of the blastocyst is a cluster of cells. This colony of **blastomeres** is the **inner cell mass (ICM)** which will give rise to the embryo proper.

We are now at the **end of the first week of human development** (keep track of gestation age!), and implantation of the blastocyst into the uterine wall has begun (Fig. 3.3 a). The trophoblastic cells over the ICM make contact and fuse with cells of the **endometrium** lining the uterine cavity. Microscopic examination of the trophoblast reveals its early differentiation into an inner, cellular layer, the **cytotrophoblast**; and an outer amorphous, syncytial layer—the **syntrophoblast**. It is the latter which advances ever more deeply into maternal tissue, the ICM being situated at its invasive edge. By day 12 of gestation (Fig. 3.3 b), the blastocyst is fully embedded in the uterine lining—and the woman may not even realize that she is pregnant! The endometrial cells proliferate, enlarge, and accumulate fat and glycogen, making the uterus a rich supply house of nutrients even before the placenta takes shape. Uterine glands become active and distend the tissues with their secretory products (uterine milk). This highly vascular and edematous endometrium of pregnancy is now renamed **decidua**.

With the blastocyst deeply implanted, usually high up on the posterior wall of the uterus, three distinct regions of decidua are defined (Fig. 3.4): **decidua basalis**, the area of decidua in contact with the embryonic pole of the blastocyst ("polar" trophoblast and ICM),and the future maternal contribution to the placenta; **decidua capsularis**, decidual tissue that covers all aspects of the blastocyst except the embryonic pole and separates the submerged blastocyst from the uterine cavity; and **decidua parietalis**, the rest of the decidua.

Placentation

The placenta is the organ which eventually assumes all functions that sustain life *in utero*. The placenta begins to form towards the third week of pregnancy, but it continues to grow and mature in step with the growth and development of the embryo and fetus. The trophoblast over the inner cell mass and the decidua opposite it contribute to the establishment of the placenta, but maternal and fetal bloodstreams never mix; they are anatomically separated by the **placental barrier**, mainly syncytial trophoblast and fetal endothelium, across which two-directional physiologic exchange takes place.

Embryogenesis

The ICM gives origin to the embryo proper. At first a solid body of cells projecting into the segmentation cavity, the ICM is transformed as the first week of human development ends. The blastomeres drift apart and undergo rearrangement, a process that creates two cavities: one is the future **amniotic cavity**, the other the primitive **yolk sac** (Fig. 3.3 c–d). They are separated by a row of tall columnar cells, the **epiblast** (known as the *formative cell layer*), resting upon a second layer of cells, the **hypoblast**. This bilaminar plate is the **embryonic disc**. Saucer-shaped at the time when the blastocyst is implanting in the decidua, on or about day seven post-conception, it will elongate and acquire polarity over the next few days. At the close of the second week of **development** (day 14 following conception), the bilaminar embryo has cranial and caudal ends. Epiblast, the floor of the amniotic cavity, forms its dorsal surface; and hypoblast, the roof of the yolk sac, its ventral surface. The rest of the yolk sac is made up of a thin layer of cells, **Heuser's membrane**, whose origin is poorly understood.

The Conceptus

This may be the appropriate time to introduce the term *conceptus*, which refers to the embryo *and* its fetal membranes. **Fetal membranes** are extra-embryonic support structures. Though not actively involved in laying down tissues and organs inside the embryo, all make important contributions to embryonic and fetal development. Both the amnion, *i.e.*, the roof and sides of the amniotic cavity, and the yolk sac are classified as fetal membranes. The other two fetal membranes are the chorion and the allantois.

At the end of the second week of human development the conceptus is roughly 3 mm in diameter, though the embryo itself is only about 0.2 mm in length. Just one week later, however, the conceptus measures approximately 1 cm and the embryo 1–2 mm.

The Trilaminar Embryo

Embryonic mesoblast is the third and last cell layer to appear; and, once in place, it converts the bilaminar embryonic disc into the trilaminar embryo.

Embryonic mesoblast appears early in the third week of development (days 15–21 of gestation), when the cells of epiblast near the hind end of the pear-shaped embryonic plate proliferate and crawl towards the midline (Fig. 3.5). In this position, the cells elongate, migrate below, and lose continuity with the surface cells (Fig. 3.6). Having gained the interval between epiblast and hypoblast, the cells of this lineage, now designated embryonic mesoblast, spread in all directions to the margins of the disc. Near the head end of the embryo, however, in the midline, embryonic mesoblast fails to penetrate tightly joined cells of epiblast and hypoblast (Fig. 3.7). This is the area of the **prechordal plate**, the future oropharyngeal membrane. Advancing columns of embryonic mesoblast will skirt the edges of the prechordal plate, then reunite in front of it. The mass of tissue at the extreme rostral pole of the embryonic plate is the **septum transversum**; between it and the prechordal plate is the **cardiogenic area**.

The midline thickening in the caudal region of the embryo, produced by dorsal surface cells pushing inwards, is the **primitive streak**. At is extreme cephalic end is another site of intense surface cell proliferation, the **primitive node**, whose center is the **primitive pit**. Cells of epiblast migrating into the gap between epiblast and hypoblast through the primitive pit advance towards the head end of the embryo, strictly in the median plane, until their forward progress is arrested by the prechordal plate (Fig. 3.7). This secondary mesoblast of primitive pit origin is so-called **chorda mesoderm**, and the rod of cells it produces is the **notochord**—the embryo's central axis. Other dorsal surface cells pouring through the primitive pit and upper primitive streak infiltrate the hypoblast and establish themselves as **embryonic endoderm**. It is now proper to refer to the cells of the epiblast as **embryonic ectoderm** and to those of the intermediate layer as **embryonic mesoderm**.

We are now at the **end of the third week of human development** (ca. day 21), the three germ layers are in place, and organ development is about to begin.

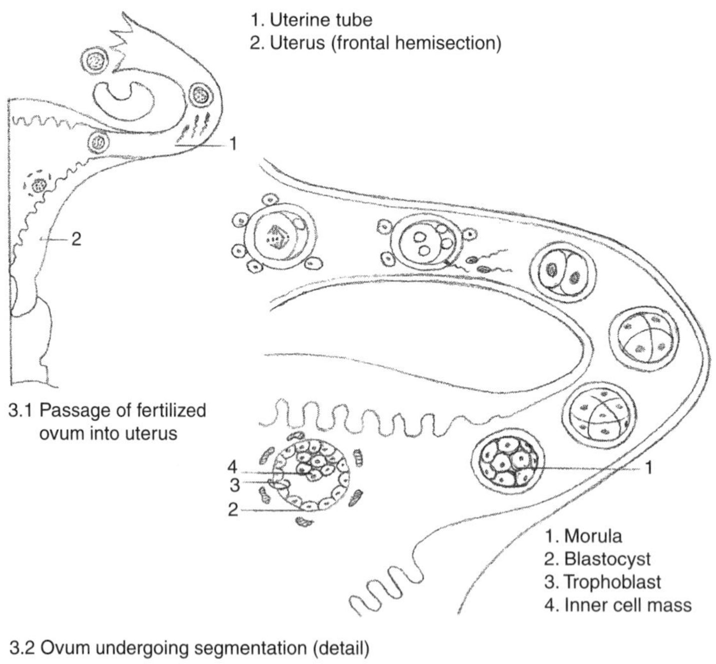

Figure 3.1: Passage of fertilized ovum into uterus
Figure 3.2: Ovum undergoing segmentation

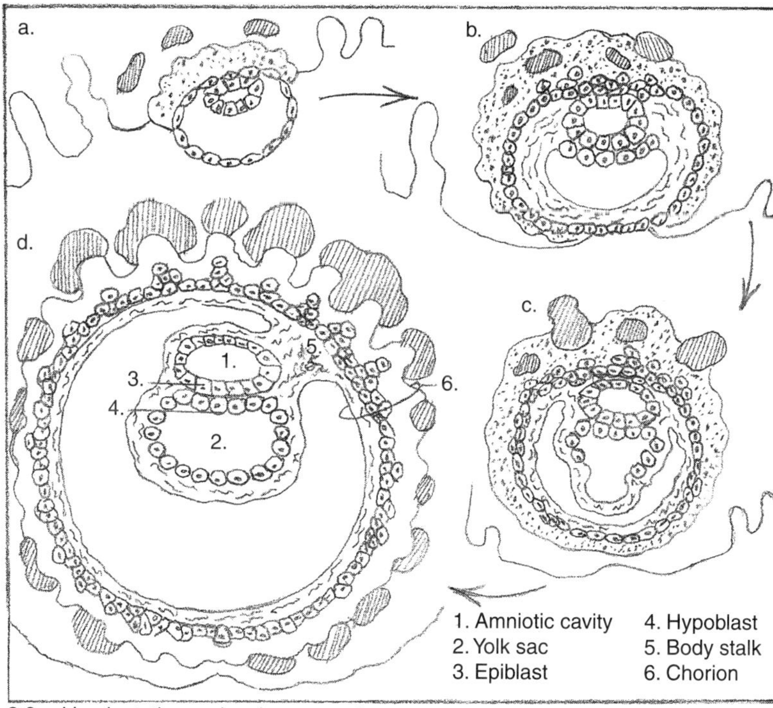

3.3a-d Implantation and embryogenesis

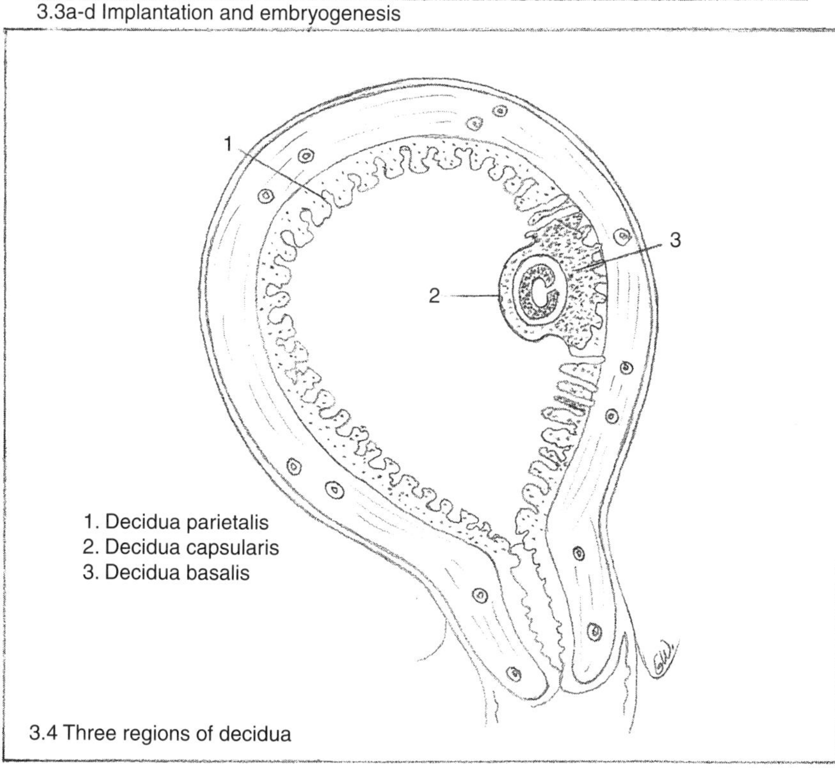

3.4 Three regions of decidua

Figure 3.3: Implantation and embryogenesis
Figure 3.4: Three regions of decidua

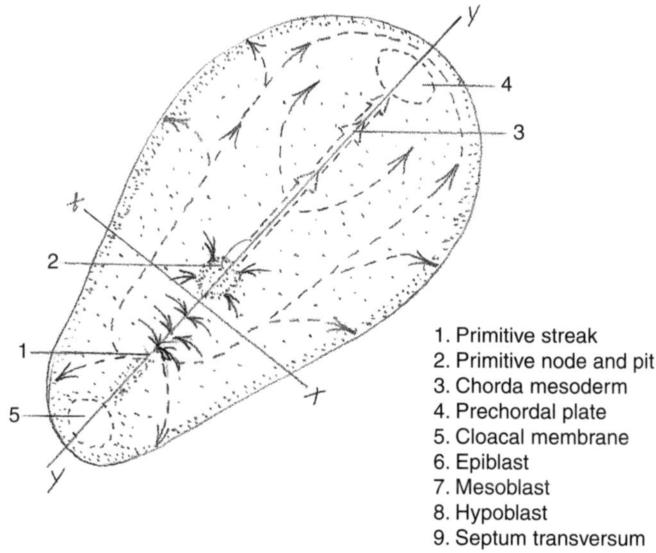

3.5 Embryonic disc and migration of epiblast cells

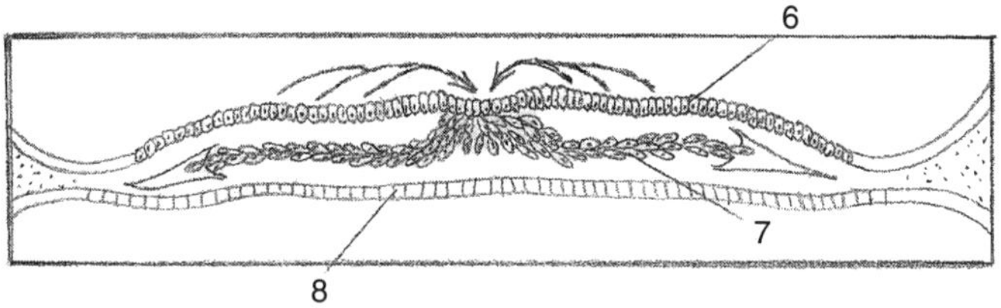

3.6 Section through embryonic disc, plane x-x

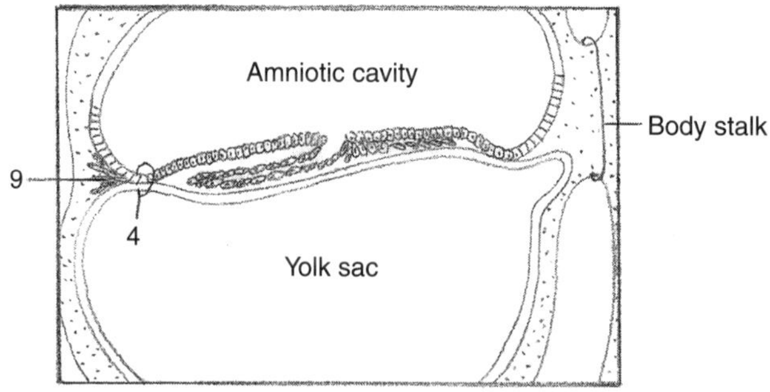

3.7 Section through embryonic disc, plane y-y

Figure 3.5: Embryonic disc and migration of epiblast cells
Figure 3.6: Section through embryonic disc, plane x-x
Figure 3.7: Section through embryonic disc, plane y-y

Chapter 4
ORGANOGENESIS

The fertilized ovum, a single cell, over time develops into a complex organism composed of trillions of cells of differerent sizes, shapes, and functions.

What is perhaps most amazing in the scheme of things is the speed with which all major organ systems of the body are laid down. It is hard to imagine, though a fact, that by the end of the sixth week following conception, the embryo has acquired external features and characteristics that stamp it unmistakably human.

While there is rapid development of embryonic form and function, incremental growth of the embryo is relatively slow at first. Its size during the first few weeks is in the millimeter range; because of its small size, however, the embryo is assured of adequate nutrition by diffusion before the placenta becomes fully functional.

The "Critical Period" in Organogenesis

Keep in mind that conception usually takes place two weeks before the woman is expecting onset of menstrual bleeding. By the time she has missed her monthly period and suspects that she is pregnant, embryonic development is well underway. Pregnancies that go unrecognized for the first two or more weeks put the developing human at risk for birth defects, because the pregnant woman may unwittingly expose the embryo to biohazards at a time when it is most susceptible to them. The **critical period** of human prenatal development begins towards the end of the third week of embryonic life, coincidental with onset of organogenesis, and ends about one month later—by which time most organs have assumed their definitive form. This is a period of active cell replication and differentiation—and a window of vulnerability, because actively dividing cells are more plastic or pliable than resting cells. Environmental agents, such as irradiation, that disturb cell division or interfere with cell differentiation during this formative stage are likely to produce malformations in the unborn. Maternal exposure to the same hazards may have no adverse physical effects on the embryo once the critical period has passed.

The Presomite Embryo

At gestation age 2+1/2 weeks, the embryo is about 1mm in length (Fig. 4.1). It is trilaminar and pear-shaped, its narrower tail end being attached to the chorion by the body stalk. Embryonic mesoderm separates ectoderm dorsally from endoderm ventrally, except in the region of the prechordal plate where ectoderm and endoderm are in direct contact. A similar contact area of ectoderm and endoderm exists near the embryo's hind end; this is the future cloacal membrane.

The embryonic mesoderm which has spread to the margins of the embryonic disc now undergoes regional modification, following which three, bilaterally symmetrical, distinct zones of mesoderm can be recognized (Fig. 4.2). They are, from medial to lateral: paraxial mesoderm, intermediate mesoderm, and lateral plate mesoderm. **Paraxial mesoderm**, as the name suggests, is situated closest to the embryo's central axis—the notochord; and, like it, paraxial mesoderm stops short of the prechordal plate. It is primarily involved in the formation of the axial skeleton and related skeletal muscle. **Intermediate mesoderm** is the source of urogenital organs, and the subject of future discussions. **Lateral plate mesoderm** extends to the margins of the embryonic disc, including its cephalic edge, and is associated with the development of the cardiovascular system and of body cavities—pericardial, pleural, and peritoneal.

The Somite Embryo

As the third week of human development ends, the paired columns of paraxial mesoderm begin to break up. Each becomes transversely segmented into blocks of tissue, or **somites** (Fig. 4.3); and, when the process is complete, towards the end of the fifth week, 40+ pairs of somites will have been created. Intermediate mesoderm also will have become segmented, but that's another story.

Somite development starts in the head region and progresses systematically in a caudal direction and a time-dependent manner, so that the number of somites present may be used as an index of the embryo's gestation age. Thus, a human embryo with seven pairs of somites is roughly three weeks old (and some 2 mm long), and one having 25 pairs is close to four weeks of age (and 3–4 mm in length).

Somites are arranged segmentally and named according to the region they occupy in the body. The first four pairs in the future head region are designated occipital somites. The eight pairs in the future neck are cervical somites. There are 12 pairs of thoracic somites in the chest region, and five pairs of lumbar and sacral somites, respectively, in the future abdomen and pelvis. Ten pairs of coccygeal somites are located in the never developing tail region. With respect to the occipital somites, the first pair disappears; the remaining three pairs contribute to the base of the skull—though no trace of its segmental origin is evident at birth. With the eventual disappearance of the terminal five paired coccygeal somites, only 38 pairs of somites will be available to the embryo as building blocks for the development of the axial skeleton and related muscle. I shall take up their subsequent differentiation and fates in chapter 5.

Types of Embryonic Tissue

Embryonic ectoderm, mesoderm, and endoderm will generate all cells, tissues, and organs of the body. A tissue is a collection of similar cells, and tissues combine to form organs.

Based on type and texture, the cells and tissues of the early embryo are classified as either epithelial or mesenchymal.

Epithelial tissue is composed of cells that typically form uninterrupted sheets or dense masses. The cells are connected by gap or tight junctions; consequently, there is little intercellular matrix in epithelial tissue. Ectoderm and endoderm, in particular, give rise to epithelia, and most organs derived from these germ layers retain their epithelial character. However, embryonic mesoderm also may form epithelia at certain times or in certain places. For instance, all blood vessels and body cavities are lined internally with a single row of flat cells, clearly mesodermal in origin, but epithelial in form. Likewise, many mesodermal structures, such as the somites, begin as epithelia, pass through a mesenchymal phase, and recondense into epithelial tissue.

Mesenchymal tissue, or **mesenchyme**, is made up of star-shaped (stellate) cells, loosely packed, hence separated by variable amounts of intercellular substance. Mesenchyme, rich in fibers, forms the basis of embryonic connective tissue. Its cells are capable of ameboid motion and often are phagocytic. Moreover, they are primitive or undifferentiated (unprogrammed) cells that have the potential of evolving into more specialized forms. The major source of mesenchyme is embryonic mesoderm, but the other two germ layers may contribute to it—and do, especially in the head region.

Both epithelial and mesenchymal tissues participate in the morphogenetic process: changes in cell size, shape, and number, as well as modifications in the adhesive properties and migratory behavior of cells, affect the form and dimensions of the embryo. Furthermore, interactions between epithelial and mesenchymal tissues are known to influence embryonic growth and differentiation.

Endoderm is the germ layer precursor for all parts of the digestive system. Outgrowths from the primitive digestive tube, such as the liver and pancreas, the thyroid gland, even the lungs, likewise are derivatives of embryonic endoderm. Ectoderm, on the other hand, is the foundation of the nervous system. I would like to discuss its beginnings briefly, because the development of this system provides insight into the basic kinetics of organogenesis.

Neurulation

Nervous system development begins late in the third week of embryonic life, when the ectoderm occupying the midline position ahead of the primitive node thickens to form an hour-glass-shaped plate of cells—the **neural plate** (Fig. 4.4). The prevailing view is that chorda mesoderm exerts an organizing or inductive influence on the ectoderm overlying it.

The plate of neurectoderm soon develops a median furrow, the **neural groove**, flanked by **neural folds** (Fig. 4.5 a–d). They move towards each other, meet, and fuse in the midline, thus converting the neural groove into the **neural tube**. The tube pinches off from the surface, beginning in the embryo's midsection and, by the end of the fourth week of development, lies free inside embryonic mesoderm. It will form the central nervous system (CNS), *i.e.*, brain and spinal cord.

Neural Crest Tissue

Neural crest tissue refers to neurectodermal nests of pluripotent cells located early on in the wings of the neural folds. As the neural tube detaches from surface ectoderm, so do neural crest cells. They arrange themselves bilaterally and segmentally in the plane of the developing somites, dorsolaterally to the neural tube. In this position, neural crest cells differentiate into sensory neurons, whose cell bodies form spinal (dorsal root) and cranial sensory ganglia; *i.e.*, they become nerve cells that transmit signals to the CNS. Neural crest tissue, therefore, plays a major role in fashioning the peripheral nervous system (PNS). Other neural crest cells migrate ventrally to form neurons of the autonomic nervous system (ANS), including those of the sympathetic chain. In addition, migrating neural crest cells differentiate into the secretory cells of the adrenal medulla and the pigment cells of the skin (melanocytes). The pia-arachnoid membrane as well as the (Schwann) cells of the neurolemma, the supporting tissue of the PNS, also arise from neural crest tissue.

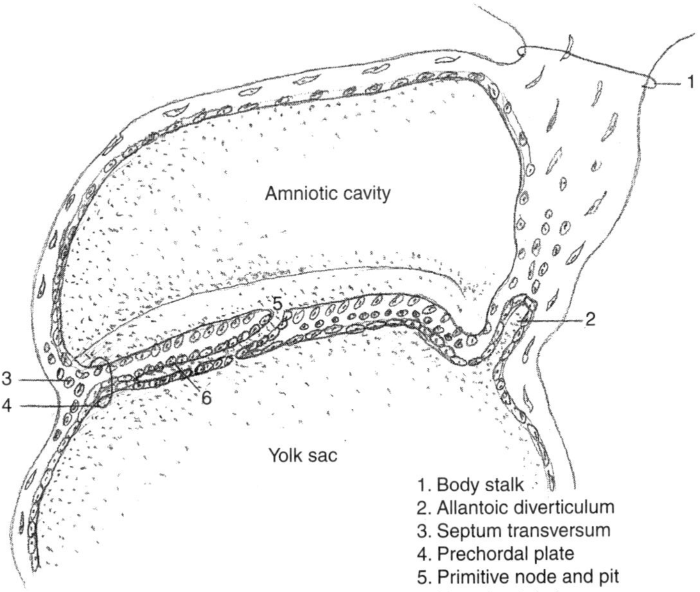

4.1 Presomite embryo (midline section)

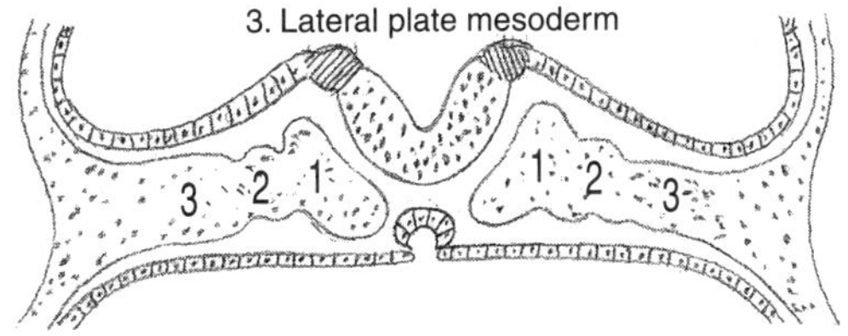

4.2 Presomite embryo (cross-section)

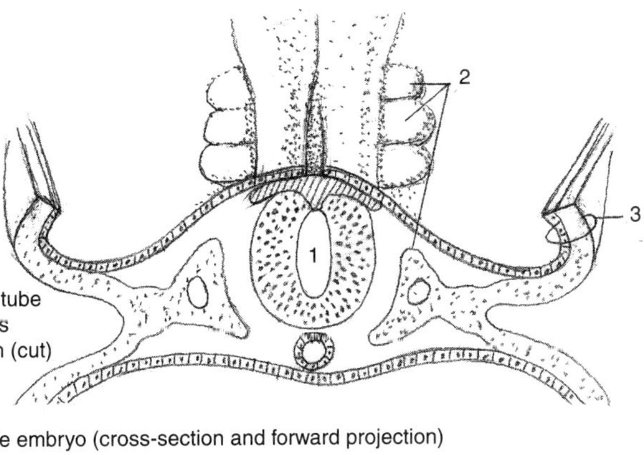

4.3 Somite embryo (cross-section and forward projection)

Figure 4.1: Presomite embryo (midline section)
Figure 4.2: Presomite embryo (cross-section)
Figure 4.3: Somite embryo (cross-section and forward projection)

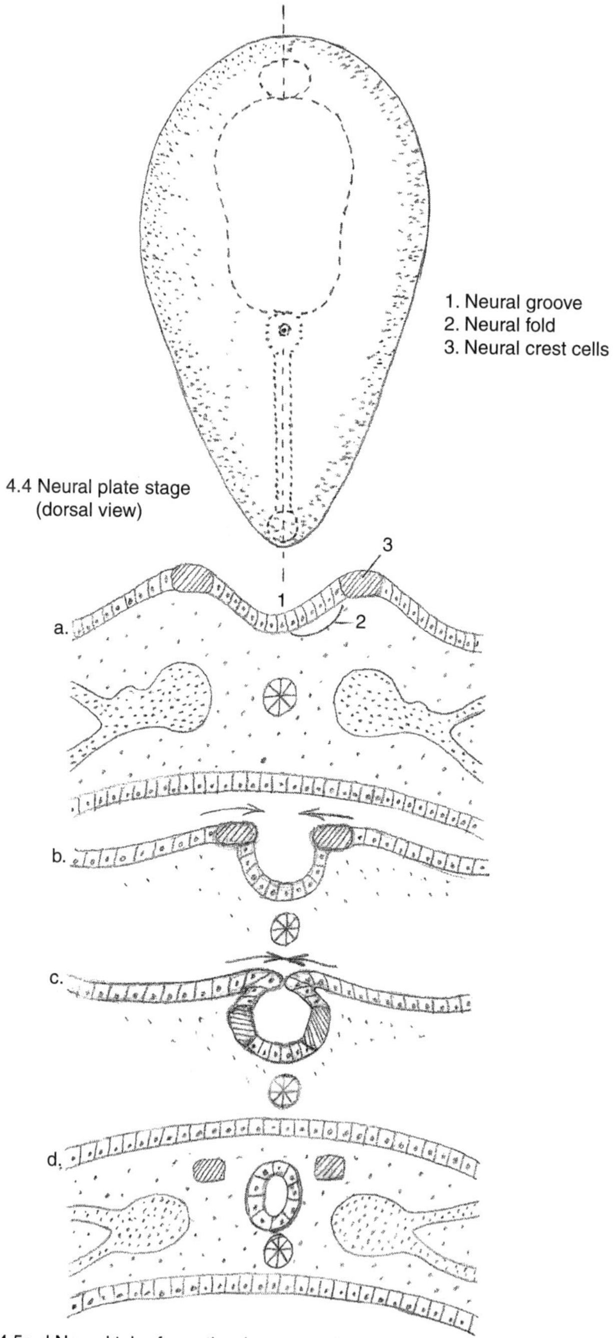

4.5a-d Neural tube formation (cross-sections)

Figure 4.4: Neural plate stage (dorsal view)
Figure 4.5a-d: Neural tube formation (cross-sections)

Chapter 5
MUSCULOSKELETAL ANATOMY

Transverse segmentation of the paraxial mesoderm late in the third week of human development imposes a pattern of periodicity, or metamerism, on the embryo. Traces of the primitive segmental organization are found in the serial repetition of vertebrae, ribs, and intercostal muscles. Also, the horizontal tendinous inscriptions of the rectus abdominis muscle remind us that it is formed by fusion of muscle primordia from several segments. Many other muscles, such as the deep muscles of the back, arise in similar fashion—though, their sheer bulk largely obscures their segmental origin. Recall that, as discussed in chapter 4, even primitive neural crest tissue is distributed in a segmental manner, an arrangement which accounts for the regular, sequential emergence of 30+ spinal nerves from the spinal cord of the adult.

Differentiation of Somites

By gestation-age 26 days, *i.e.*, approximately one week after somites first appear (see chapter 4), changes have occurred that cause each to appear tripartite. The ventromedial part of each somite is the **sclerotome**, the dorsomedial part is the **myotome**, and the lateral part is the **dermatome** (Fig. 5.1).

Sclerotoginous cells differentiate into the hard and soft tissues of the axial skeleton, *e.g.*, bones, cartilage, and ligaments of the vertebral column, ribs, and sternum. They also are the source of the spinal dura mater, the outer, tough covering of the spinal cord. Myotomes, meanwhile, seed myoblasts, the precursor cells of skeletal muscle, their tendons, and connective tissue. Be aware, however, that only about 270 of the 600+ muscles of the body originate from paraxial somites; the limb muscles have a different developmental history, as do the muscles of the head. The cells of dermatomes are programmed to lay down dermis, the connective tissue layer of the skin.

The Vertebral Column

The vertebral column, or "spine," is composed of some 33 vertebrae stacked vertically in single file, and tied together by intervertebral discs and ligaments which convert this column of bones into a stiff yet flexible support structure (Fig. 5.2). Its average height is 70 cm (28 in.) in the male and 60 cm

(24 in.) in the female; one-fourth to one-fifth of the column length is due to the presence of intervertebral discs. We shrink as we age, because discs lose resiliency and flatten.

The vertebrae are classified according to their location in the body. The seven in the neck region make up the cervical (C) collection. There are 12 vertebrae in the thoracic (T) region, and five each in the lumbar (L) and sacral (S) regions. Three to five vertebrae compose the coccygeal (Cc) set. The five sacral vertebrae are fused to form the sacrum, and fusion converts the coccygeal vertebrae into the coccyx.

Vertebrae of one region have one or more distinctive features that set them apart from those of other regions. All vertebrae, however, have the same basic form (Fig. 5.3): a kidney-shaped body and a vertebral (neural) arch. Only C1, the atlas, and C2, the axis, are atypical in that respect (Fig. 5.4).

Vertebral bodies transfer the weight of the head and torso to the hip joints and lower extremities. Consequently, bodies increase in size progressively from top to bottom as the vertical load increases (Fig. 5.2). Each vertebral arch encloses a triangular space, the vertebral foramen. Collectively, when laid end to end, vertebral foramina (L., plural of *foramen* = a passage or an opening) form the vertebral canal, occupied in life by the spinal cord which, therefore, is safely encased in bone. Projecting from each vertebral arch are transverse and spinous processes that serve to attach muscles and function as lever arms increasing the mechanical advantage of muscles moving the vertebrae. Vertebral arches of contiguous vertebrae form interlocking zygapophysial (vertebral arch) joints (Fig. 5.5). Coupling is achieved through paired articular processes—the inferior processes of the vertebra above articulating with the superior processes of the vertebra below. The joints help create intervertebral foramina that transmit spinal nerves.

Intervertebral Discs

Adjoining vertebral bodies are united by intervertebral discs, each composed of concentric rings of fibrocartilage externally, the **anulus fibrosus**, and a gelatinous mass internally—the **nucleus pulposus**, a remnant of the embryo's notochord (Fig. 5.6).

Being softer than bone, intervertebral discs allow independent movement between and among vertebrae; and, being pliable and capable of altering their shapes, intervertebral discs dampen the impact of sudden stresses and strains. Like the vertebral bodies which they unite, intervertebral discs are subject to vertical compression forces that become maximal on the last two lumbar vertebrae and discs. Under the strain, the nucleus pulposus may herniate through the anulus into the vertebral canal. A "slipped" or prolapsed disc, as this condition is known clinically, often puts pressure on the spinal cord or nerve roots, causing lower-back ache and shooting pains down the leg.

Ligaments of the Vertebral Column

While intervertebral discs unite neighboring vertebrae to provide a degree of flexibility to the spine, ligaments exist to maintain the structural integrity of the column without interfering with its movements. Some, like the anterior and posterior **longitudinal ligaments**, are densely fibrous and stretch-resistant (Fig. 5.7). Others, such as the **ligamenta flava** (L., plural of *flavus* = yellow), are made of elastic tissue. Here, then, is an important lesson in functional anatomy: vertebral bodies and intervertebral discs, both of which are primarily weight-bearing structures, are tied together by strong and unyielding longitudinal ligaments to prevent destabilization of the spine by gravitational forces. In contrast, adjoining vertebral arches are connected to each other by elastic ligamenta flava that will stretch when muscles, acting through vertebral arches, flex, extend, or rotate the vertebral column.

Muscles of the Axial Skeleton

The muscles of the body come in three varieties: somatic, visceral, and cardiac. All trace their origin to the same embryonic germ layer—mesoderm.

The muscles of the axial skeleton, including those that move the vertebral column, are distinctly somatic: *i.e.*, each is composed of multinucleated, striated fibers innervated by somatic motor nerves, placing them under voluntary control. The source of somatic muscle is paraxial mesoderm, specifically the part of the somite known as myotome. In embryos of gestation ages five or six weeks, primitive myotomes are still visible. Each splits into dorso- and ventrolateral parts opposite the transverse process of a vertebra, a division, which foreshadows the formation of pre- and postvertebral muscles. By the end of the second month of intra-uterine life most of the adult muscles can be recognized and, by that time, have acquired their contractile powers.

Functionally, as their location dictates, prevertebral muscles are flexors of the spine, and postvertebral muscles are extensors. Prevertebral muscles are prominent only in the cervical and lumbar regions, where they may act to straighten the anteriorly convex curvatures of the vertebral column. Postvertebral muscles are more massive and more powerful than their opposites, because in the upright posture they must oppose gravitational pull, which tends to force the body into forward flexion. In other words, postvertebral muscles help keep the spine erect; they are members of the

erector spinae group, the deep muscles of the back. The head also will flex on the chest unless postvertebral neck muscles, such as those of the suboccipital group, remain vigilant. If unsupported, the head will fall forwards, because its center of gravity is in front of the atlanto-occipital joint formed by C1 and the occipital condyles at the base of the skull. The **ligamentum nuchae** (Fig. 5.8), the upward continuation of the supraspinous ligament, crosses the joint and helps stabilize the head in a neutral position. It is stretch-resistant in the human, but composed of elastic tissue in long-necked animals.

Gravity and Musculoskeletal Anatomy

It must be obvious from the foregoing discussion that the pull of gravity is an external force capable of influencing the musculoskeletal machinery of the body. It imparts acceleration to an object, such as the head, unless opposed by an equal but opposite force—such as postvertebral neck muscles. In the absence of antigravity muscles, the knee would buckle and the ankle collapse, causing us to land on the floor. The muscles of the anterior thigh, the quadriceps group, and the muscles forming the calf of the leg maintain erect posture by counteracting gravitational stress in a sustained manner. No wonder the front of the thigh, the calf of the leg, and the back of the trunk are bulkier and more massive than the reciprocal surfaces.

But gravitational force can be help as well as hindrance. Here are two examples: In the course of inspiration, during the respiratory cycle, the thorax is raised by muscular action. Though muscles exist to lower it during the expiratory phase, the thoracic cage is returned to its resting position, effortlessly, by gravitational pull. Likewise, following abduction of limbs from a body in upright posture, adduction is a passive process achieved by force of gravity. Powerful adductor muscles, such as pectoralis major or adductor magnus, are not required to act unless adduction has to be executed against resistance. Such is the case, of course, in gymnasts pulling themselves up on horizontal bars, or in swimmers doing the breast stroke.

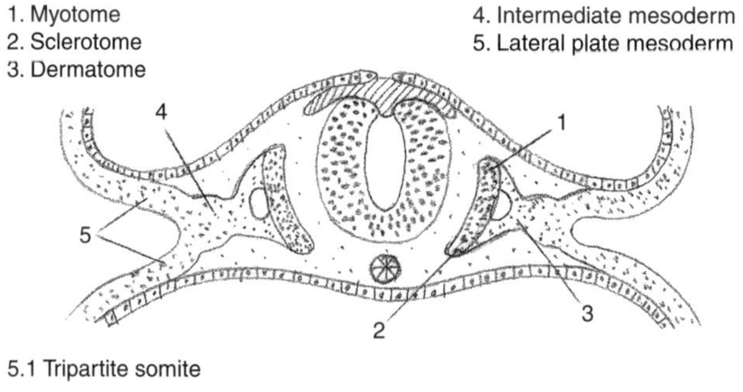

Figure 5.1: Tripartite somite

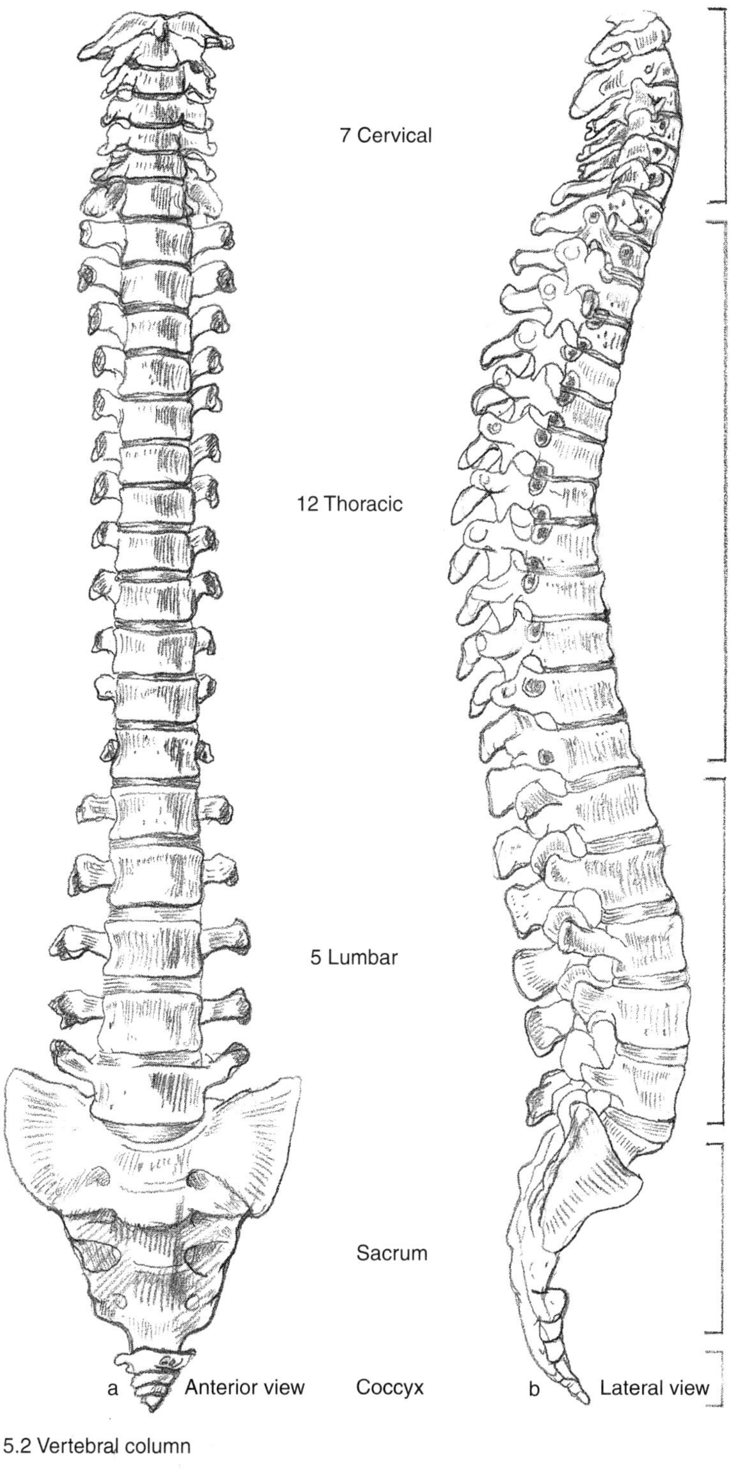

5.2 Vertebral column

Figure 5.2: Vertebral column anterior and lateral views

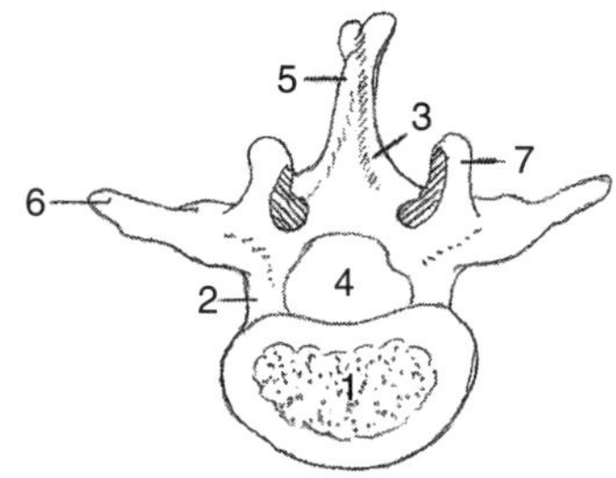

1. Body
2. Pedicle
3. Lamina
4. Vertebral foramen
5. Spinous process
6. Transverse process
7. Articular process

5.3 Typical vertebra

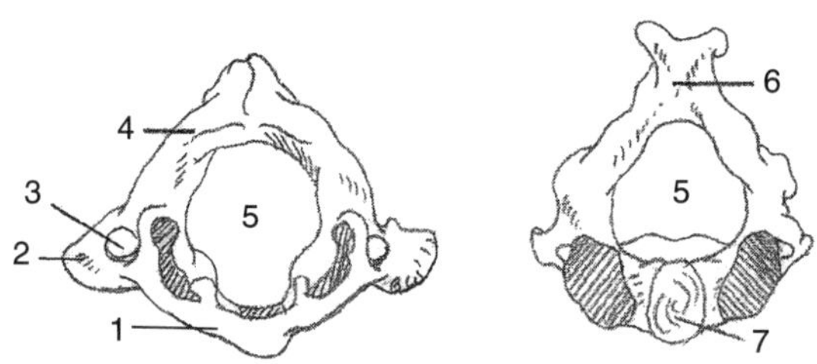

1. Anterior arch
2. Lateral mass
3. Foramen transversarium
4. Posterior arch
5. Vertebral foramen
6. Spinous process
7. Dens

5.4 Atlas and axis

1. Superior articular process
2. Inferior articular process
3. Intervertebral foramen

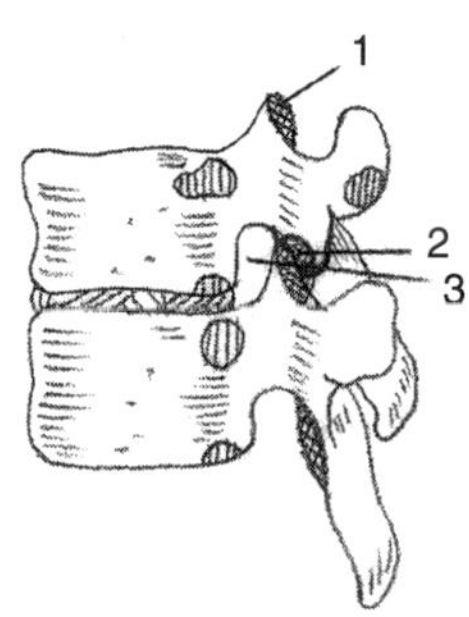

5.5 Vertebral arch joints (lateral view)

Figure 5.3: Typical vertebra
Figure 5.4: Atlas and axis
Figure 5.5: Vertebral arch joints (lateral view)

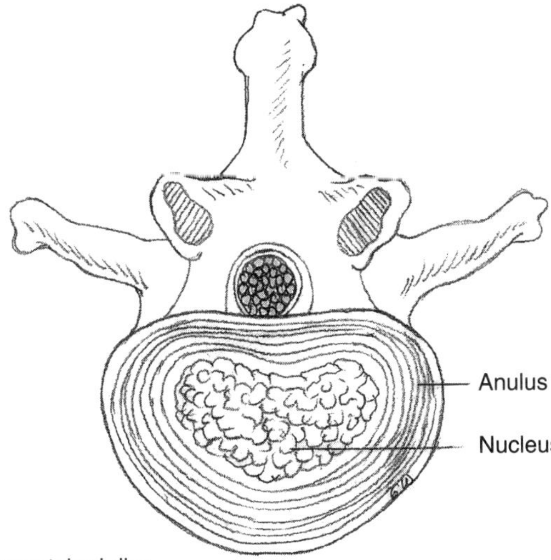

5.6 An intervertebral disc

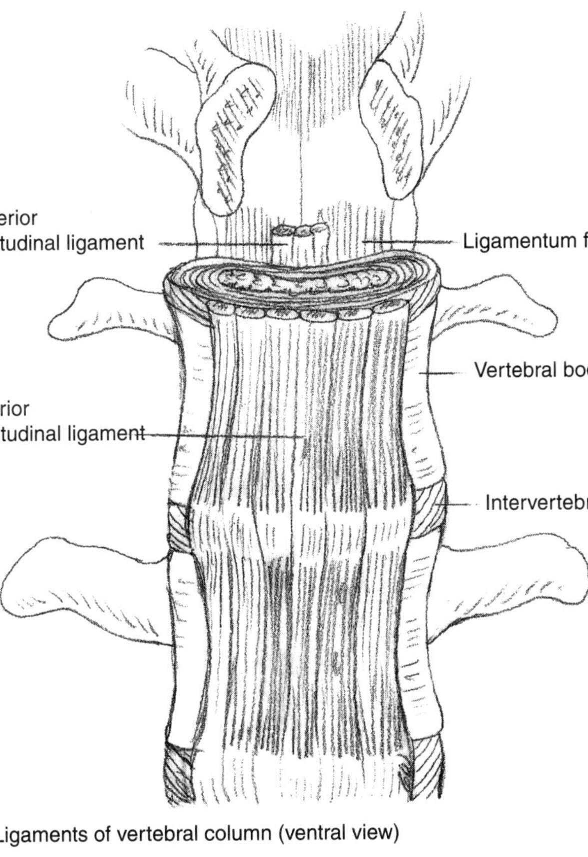

5.7 Ligaments of vertebral column (ventral view)

Figure 5.6: An intervertebral disc
Figure 5.7: Ligaments of vertebral column (ventral view)

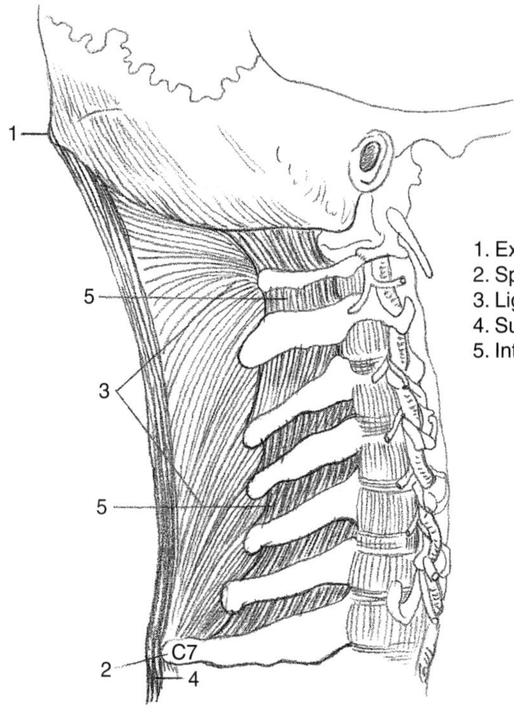

1. External occipital protuberance
2. Spinous process of C7
3. Ligamentum nuchae
4. Supraspinous ligament
5. Interspinous ligament

5.8 Ligamentum nuchae

Figure 5.8: Ligamentum nuchae

Chapter 6
THE MEMBRANES OF THE BODY

Membranes forming cell walls or lining cavities typically are complex systems mediating transfer between compartments—or, as in the case of the skin, between external and internal environments. I shall consider some of these briefly, but the main emphasis in this chapter will be on fascia which, according to many anatomists, is sufficiently important a tissue to justify its inclusion into the category of organ systems.

Skin

The skin is a compound membrane, having both epithelial and fibrous components (Fig. 6.1). The epithelial component is the **epidermis**—the outer, cellular layer of the skin. The surface cells have a rapid turnover rate; for, being exposed to air, they dry out, keratinize, and exfoliate. Their replacements are generated by stems cells from below. The epidermis rest upon the **dermis**, a layer of connective tissue replete with nerves, blood vessels, sweat and sebaceous glands, hair follicles, and arrector pili muscles responsible for "goose flesh." Incidentally, the Latin equivalent of skin is *cutis*—hence, cutaneous signifies "relating to skin," and subcutaneous "under the skin" or "below the skin."

Epidermis and dermis differ not only structurally but also developmentally in that the germ layer source of the former is ectoderm, while that of the latter is mesoderm—as is all of the body's connective tissue. The skin forms a protective, waterproof covering for the rest of the body, in addition to which it performs absorptive, secretory, metabolic, thermoregulatory, and neurosensory functions.

Mucous Membrane (Mucosa)

Like skin, the mucous membranes of the body are compound membranes (Fig. 6.2). The epithelial layer lines the internal surface of hollow organs such as the gut, and a connective tissue framework, known to histologists as the lamina or tunica propria, forms its supporting base. Mucous membranes differ from skin by not being exposed to the external environment; hence, they do not dry out or keratinize, though they, too, are constantly shed and replaced. The germ layer source of mucosal epithelium typically is endoderm.

Furthermore, sticky mucus, produced by cells and glands of mucous membranes, keeps their surfaces slippery. The mucosa of the nasal cavity and of upper respiratory passages has mucus-producing and ciliated cells to trap foreign particulate matter and expel it from the airways. Parts of the gastrointestinal mucosa contain secretory cells to aid in the digestive process.

Serous Membrane (Serosa)

Peritoneum, pleura, and pericardium are serous membranes. All are closed sacs, line body cavities, and are divisible into visceral and parietal components (Fig. 6.3). The **visceral** portion of a serous membrane is applied to an organ, such as the heart, and forms its external layer. The **parietal** portion of a serous membrane is continuous with the visceral portion beyond the limits of the organ and lines the body cavity which accomodates it. In this specific case, the body cavity is the pericardial sac.

Serous membranes are compound membranes, each composed of a single row of flat cells on a bed of loose connective tissue. The cellular layer which lines the body cavity has all the characteristics of an epithelium; but, because its germ layer source is mesoderm rather than ectoderm or endoderm, we refer to the sheet of cells as **mesothelium**.

Peritoneal, pleural, and pericardial cavities contain small amounts of a low-viscocity liquid which resembles blood serum in its consistency; hence, the term serous membrane. And, because visceral and parietal components are normally separated by a thin film of serous fluid, friction between them is minimal for as long as their mesothelial surfaces remain intact.

Synovial Membrane (Synovium)

Synovial joints, *i.e.*, freely movable joints, are defined by the presence of a joint cavity filled with a small amount of viscous fluid (Fig. 6.4). The synovial fluid, which plays an important role in joint lubrication and nourishment of articular cartilage, is the product of mesothelial cells lining the joint space. The lining cell layer with its backing of connective tissue is the synovial membrane. It is applied to the inner aspect of the fibrous capsule surrounding the joint. And, as the joint capsule attaches to bone and blends with the periosteum over it, so does the synovial membrane.

In rheumatoid arthritis, a serious disease affecting joints, the synovial membrane produces large quantities of synovial fluid, causing painful distension of the joint and limitations of movement.

Superficial Fascia

We now turn from compound membranes with epithelial components to simple fibrous membranes, namely fasciae and aponeuroses.

Your first-day assignment in the gross lab probably included reflection of skin, an exercise that, when executed properly, exposed a fatty layer of connective tissue—the superficial fascia. It lies below the dermis (histologists refer to it as hypodermis) and merges with it imperceptibly.

The superficial fascia is fairly unremakable, except that it contains variable amounts of fat, superficial vessel and nerves and, in the chest region, the duct system and secretory units of the breast.

Embedded in parts of the superficial fascia are isolated patches of muscle, *e.g.*, the platysma in the neck region. They are the vestiges of the panniculus carnosus which, in certain animals, is an extensive subcutaneous muscle layer.

Deep Fascia

This membranous layer, in its various forms, is found throughout the body. You will have encountered it first beneath the skin and superfical fascia, to which it is tightly bound by stubby, ligamentous threads. The more densely they are packed, and the greater their tensile strength, the less slippage there is of skin on fascia. They are especially abundant and tenacious in the palms of the hands and soles of the feet; hence, the better the grip and the firmer the foothold. Conversely, where deep fascia is lacking, the skin is highly mobile and tends to wrinkle—as it does in the face.

In certain parts of the body, deep fascia is structurally significant and given proper names. It may, for instance, encapsulate organs such as the parotid salivary glands and acquire specific identity as *parotid fascia*. Or, it may wrap around a whole body segment such as the thigh and be labelled the *fascia lata*. Or, it may invest a neurovascular bundle—artery, vein, nerve—as it does in the neck, to become known as the *carotid sheath*.

For the most part, however, deep fascia is non-specific and structurally insignificant. Nevertheless, all deep fascia, whatever its nature, is clinically important, for it is relatively impermeable to fluids, including blood and urine, to pus, and to pathogens. Deep fascia, therefore, acts as a barrier to the spread of infection—that is the good news. But, because bacteria are unable to move out of a fascial space, they may track in its plane, thereby spreading an infection to other parts of the body in a quite predictable manner.

Deep fascia frequently is not a single sheet but laminated, like so much puffed pastry; and, in some places, it is reticulated or honeycombed. When so organized, deep fascia is likely to simplify the study of a region by dividing it into several compartments each of which can be explored independent of the rest. The multi-layered cervical fascia, *i.e.*, the deep fascia of the neck, is a case in point. The neck is a very complicated part of the human anatomy, for its many important constituents are confined to a relatively small space and difficult to sort out. The cervical fascia, by delineating smaller units of structure or function, makes the congested anatomy of the neck readily digestable.

Deep fascia is very sensitive, being supplied by the same nerves that innervate the overlying skin. Stretch triggers pain. Mumps, for instance, is painful because a swollen or inflamed parotid gland tenses the parotid fascia. Likewise, subcutaneous hemorrhages and hematomas cause pain when the accumulated blood stretches the rigid boundaries of a fascial space. Vigorous exercise also is likely to produce so-called compartment syndromes. For example, jogging may cause "shin splints," the reason being that increased blood flow to leg muscles makes them gain bulk. The swelling will tense the crural fascia, the deep fascia of the leg, and lead to pain, discomfort, even acute lameness. Finally, the condition of "Charley Horse" refers to soreness and stiffness in the quadriceps muscle of the anterior thigh. It usually is the result of a deep thigh bruise or contusion. The pain and stiffness associated with this condition also may be attributed to increased intracompartmental pressure.

Aponeuroses

Aponeuroses are connective tissue membranes that often function as flat tendons attaching muscles to bone. Being rich in stretch-resistant collagenous fibers, aponeuroses may appear white and shiny, especially when the fibers are oriented parallel to each other in the line of muscular pull.

The thoracolumbar fascia of the back may serve as an example (Fig. 6.5). Its posterior lamella is the aponeurosis connecting latissimus dorsi and serratus posterior inferior muscles to lumbar spinous processes; its glistening appearance stands in sharp contrast to the fleshy parts of those muscles.

Not all aponeuroses are as membranous or well-defined as the posterior lamella of the thoracolumbar fascia. The thick, wide, and densely fibrous palmar and plantar aponeuroses lack smooth texture and clear-cut boundaries. Located in the palm of the hand and sole of the foot, respectively, these aponeuroses not only protect deeply placed delicate tissues, but also act as tie-beams or struts for the arches of the hand and foot.

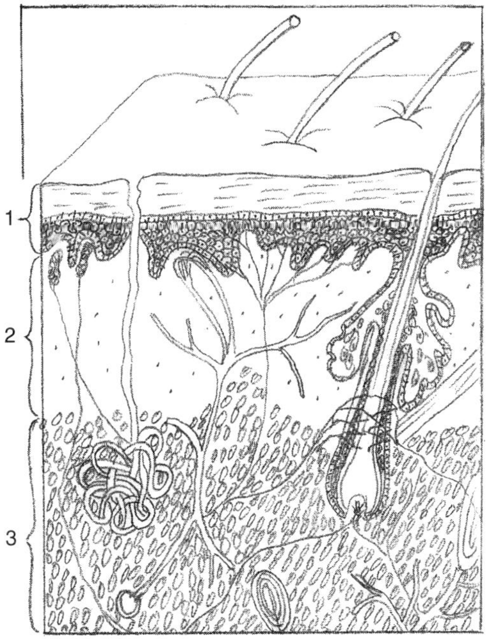

1. Epidermis
2. Dermis
3. Subcutaneous tissue

6.1 Skin as compound membrane

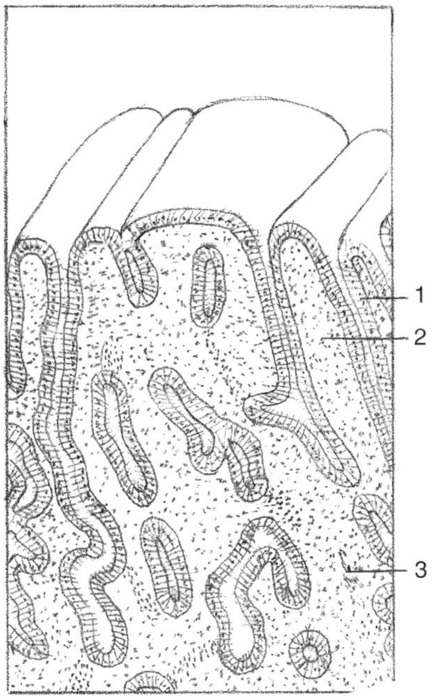

1. Epithelium
2. Lamina propria
3. Submucosa

6.2 Compound mucous membrane

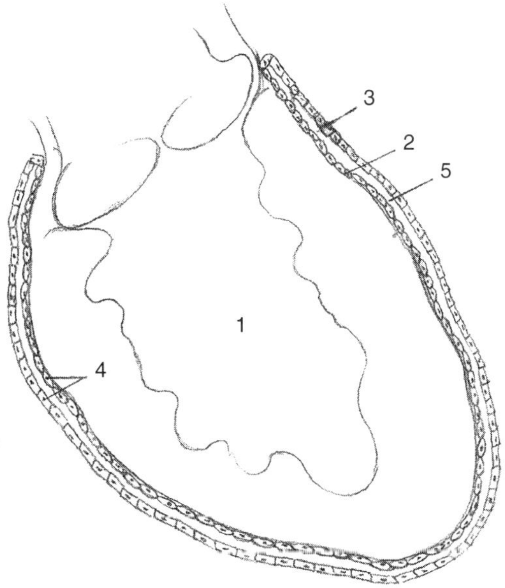

1. Heart chamber
2. Visceral layer
3. Parietal layer
4. Mesothelium
5. Serous fluid in cavity

6.3 Anatomy of serous membranes

Figure 6.1: Skin as compound membrane
Figure 6.2: Compound mucous membrane
Figure 6.3: Anatomy of serous membranes

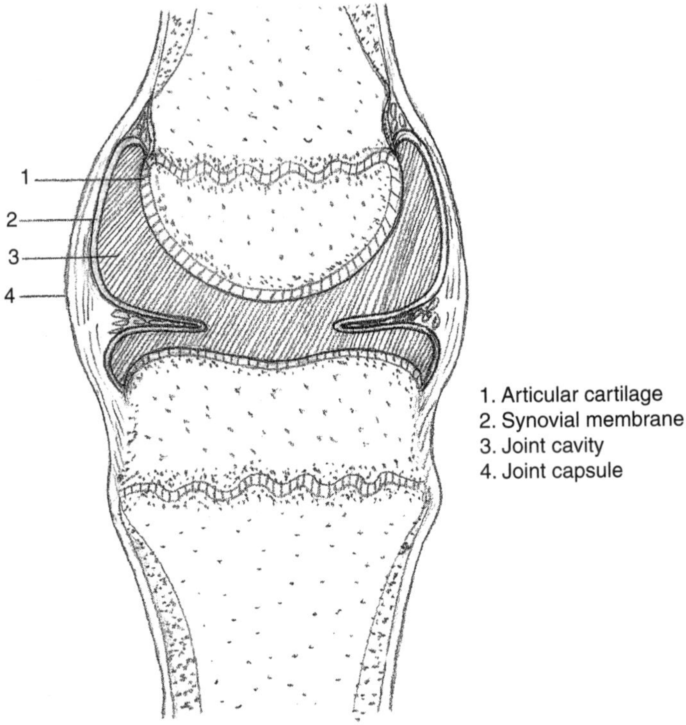

6.4 Typical synovial joint

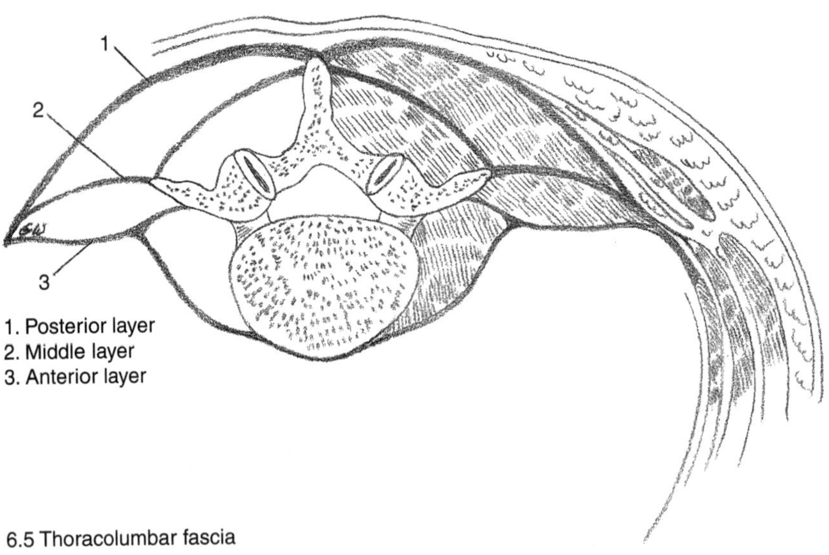

6.5 Thoracolumbar fascia

Figure 6.4: Typical synovial joint
Figure 6.5: Thoracolumbar fascia

Chapter 7
THE PECTORAL REGION

A likely etymologic source of the term "pectoral" is *pectus*, the Latin noun referring to the breast or mammary gland and, by extension, to the chest as a whole. On the other hand, *pecten* is Latin for comb; the pectoral region, then, may be so named because of the comb-like arrangement of ribs attaching to the sternum.

Myofascial Anatomy

The large, fan-shaped **pectoralis major muscle** is exposed when skin and superficial fascia of the region are reflected. The muscle is covered by **pectoral fascia**, which is prolonged below as the deep fascia of the anterior abdominal wall; in the chest region, it joins the fasciae over latissimus dorsi and serratus anterior muscles to line the **axillary fossa** (Fig. 7.1). Observe the bilaminar tendon of pectoralis major (Fig. 7.2): the fibers of the superficial lamina slant downwards, and those of the deep lamina upwards, as they pass to their insertion on the humerus. The reconfiguration of the tendon brought about by crossing laminae causes the upper fibers of the pectoralis major tendon to insert into bone distal to the lower fibers. When the arm is now abducted by muscles, such as deltoid, the tendon unwinds, thereby removing limitations on the full range of over-the-head motion. The upper and lower borders of the latissimus dorsi tendon, likewise, become reversed before inserting into the intertubercular sulcus of the humerus medial to the attachment of pectoralis major.

The **pectoralis minor muscle** lies under cover of pectoralis major. The two muscles together form the fleshy anterior axillary fold, the anterior wall of the axillary fossa. The deep fascia in the plane of pectoralis minor is the **clavipectoral fascia** (Fig. 7.3). It attaches to the clavicle above and ends below as the suspensory ligament of the axilla which, it has been said, helps maintain the concavity of the axillary floor—the arm pit. Other parts of the clavipectoral fascia enclose the subclavius muscle and pectoralis minor. The fairly dense sheet of clavipectoral fascia between these two muscles, known as the **costocoracoid membrane** or ligament, forms the floor of the infraclavicular fossa. The membrane, which shelters axillary vessels and parts of the brachial plexus, is pierced by the cephalic vein, the lateral pectoral nerve, the thoraco-acromial artery, and lymphatic vessels in route between the axilla and the deltopectoral region.

The Human Breast

The major portion of the **human breast** or **mammary gland** is embedded in the superficial fascia of the pectoral region. Thus, it lies superficial to the pectoralis major muscle and fascia.

In the human female, the breast is important functionally, cosmetically and, above all, clinically: close to 180,000 American women will be diagnosed with breast cancer this year alone, and one in four will succumb to it. And, here is another grim prediction from the National Cancer Institute: one in eight American women will develop breast cancer in her lifetime. Carcinoma of the breast is second only to lung cancer as the deadliest malignancy in the female population. You are well advised, therefore, to pay close attention to breast development, structure, function, and pathology.

Mammary glands develop during the second month of embryonic life as ingrowths from surface ectoderm—the same germ layer that gives rise also to the epidermis. And, like the epidermis, breast tissue is epithelial tissue. At first, the gland consists of a solid bud which organizes itself into 12 or more units, the future lobes of the breast. In the course of the last trimester, the lobes divide into secondary and tertiary lobules. Gradual loosening of the glandular epithelium establishes the breast's duct system; and, when fully developed, each lobe has a **lactiferous duct** fed by a network of lobular ducts. Before reaching the body's surface, the lactiferous ducts dilate to create a sinus or ampulla whose purpose, however, is poorly understood. True nipples form postnatally, but the breast lies dormant until puberty, when it increases in mass and size under hormonal stimulus. The epidermis surrounding the **nipple** becomes the pigmented **areola**.

The breast is divisible into four quadrants (Fig. 7.4). The upper outer quadrant (quadrant I) includes the **axillary tail** (of Spence)—breast tissue that extends into the axilla. Quadrant I, therefore, contains more breast tissue than other quadrants and has the longest duct system. It also has the most extensive lymph-collecting and dispersal system of any quadrant: approximately 75 percent of breast lymph passes through quadrant I to axillary lymph nodes. Lymph from quadrants II, III, and IV normally drains first into groups of nodes nearest to them.

White connective tissue incompletely separates the lobes of the breast internally and is the source of large numbers of collagenous fibers—the **suspensory ligaments of the breast** or (Cooper's ligaments). They are important, because they shape and support the breast. Breasts begin to sag when these non-elastic ligaments are torn or stretched beyond their normal limits.

Very little glandular epithelium is present in the breast at puberty or, for that matter, in the non-pregnant female. The postadolescent breast exists mainly as a system of ducts dispersed in superficial fascia and submerged in variable amounts of fat. In women who have conceived, however, additional lobules develop early in pregnancy; from the tips of ducts cells bud off to form **alveoli**, the secretory elements of the breast. Bunched like grapes on the vine, their growth is dependent on the presence of a placental hormone, **human placental lactogen** (HPL). Then, after the baby is born, maternal **prolactin** (PRL) from the anterior pituitary gland stimulates milk production by alveolar cells. Milk ejection is due to the action of yet another hormone, **oxytocin**, secreted by the posterior pituitary gland. Its release is triggered by the suckling reflex; hence, whenever breast surgery is indicated, every effort must be made to preserve the nerve supply to the nipple.

Any discussion concerning breast cancer must begin with an analysis of lymph flow and drainage in the pectoral region. Most of the lymph draining from the chest wall passes to the axillary lymph nodes of the same (ipsilateral) side. Actually, there are several groups of axillary nodes, most related to the axillary vein and lateral thoracic and subscapular branches of the axillary artery. The few lymph nodes located in the clavipectoral (deltopectoral) triangle, at the point where the cephalic vein pierces the coracoclavicular membrane, receive lymph also from the upper limb. A smaller volume of lymph, mostly from the medial parts of the chest wall, is brought to a group of parasternal nodes forming a chain along the internal thoracic artery. In general, lymph vessels carrying lymph to regional nodes accompany blood vessels. In women with breast cancer, the aim of the surgeon is to strip away all lymphatics draining the breast, and to remove all nodes in the fields of lymph drainage—for they are filters that may have trapped malignancy cells.

Lymphatics of the Breast

Lymphatic vessels, including those of the breast, are important, because they are routes by which certain cancers spread to locations far removed from the primary site.

Breast lymph passes from a web of lymphatic capillaries that surround the alveoli of the breast, the **perialveolar plexus**, to **perilobular**, then **perilobar plexuses** (*N.B.*, the plural of plexus is not plexi, nor is the plural of fetus, feti!). Thereafter, lymph enters the **periductal plexus** accompanying a lactiferous duct. It carries breast lymph to subcutaneous lymphatics under the areola—the **subareolar plexus**, continuous with lymphatic vessels of the opposite side, the anterior abdominal wall, and the root of the neck. Lymph also flows towards the deep lymphatic plexus over pectoralis major and serratus anterior muscles. Superficial and deep lymph vessels communicate by way of interlobar lymphatics that accompany Cooper's ligaments.

In general, then, breast lymph normally drains centrifugally; *i.e.*, it is routed away from the gland (Fig. 7.5). When this normal pattern of lymphatic drainage is blocked as a result of neoplastic disease, lymph accumulates in the subcutaneous tissues and stretches the skin. Cancerous invasion of lactiferous ducts and Cooper's ligaments causes them to contract which, in turn, leads to irregular dimpling of the skin that now begins to feel and look like orange peel (*peau d'orange*). In advanced stages of mammary carcinoma, there is retraction of the nipple, fixation and distortion of the breast, and reddening of the skin (hyperemia) owing to increased blood flow in the area of the neoplastic growth.

As discussed earlier, approximately 75 percent of breast lymph passes to axillary nodes through quadrant I. It, therefore, has the highest probability of becoming cancerous, and statistics indeed show the incidence of lesions to be highest in quadrant I. Malignancies originating in the upper outer quadrant, however, have a much better prognosis than those developing in the lower inner quadrant—cancer cells can spread with relative ease from quadrant IV to the contralateral breast and to abdominal organs. Fortunately, quadrant IV is least often involved in carcinoma of the breast.

Arteries of the Pectoral Region

Branches of the axillary artery supply the lateral chest wall, including the outer quadrants of the breast: superior thoracic, thoraco-acromial, and lateral thoracic arteries to quadrant I, and subscapular artery to quadrant II. The internal thoracic artery, formerly known as the internal mammary artery, the second branch off the subclavian artery, brings blood to the medial chest wall, including breast quadrants III and IV. Perforating branches of intercostal arteries, connecting thoracic aorta and internal thoracic vessels, also distribute to breast tissue. Abundant communications (anastomoses) exist between these various sets of arteries, making the breast a vascular hub through which arterial blood may be redirected in the event that normal flow through either subclavian or axillary vessels is obstructed.

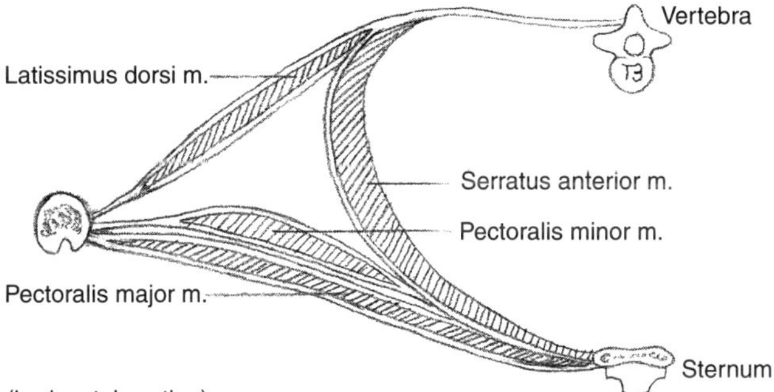

7.1 Axilla (horizontal section)

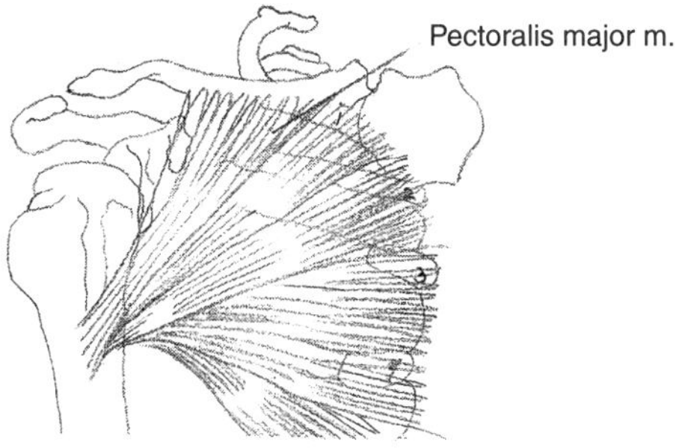

7.2 Bilaminar pectoralis major tendon

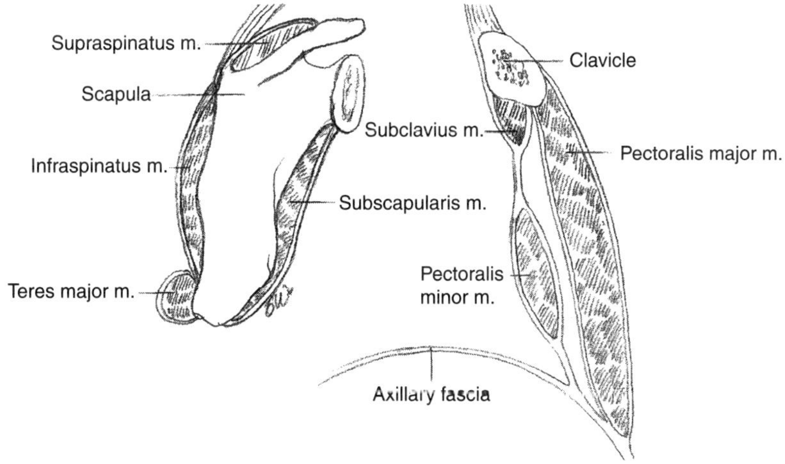

7.3 Axilla (vertical section)

Figure 7.1: Boundaries of the axillary fossa
Figure 7.2: The bilaminar tendon of pectoralis major
Figure 7.3: Axilla (vertical section)

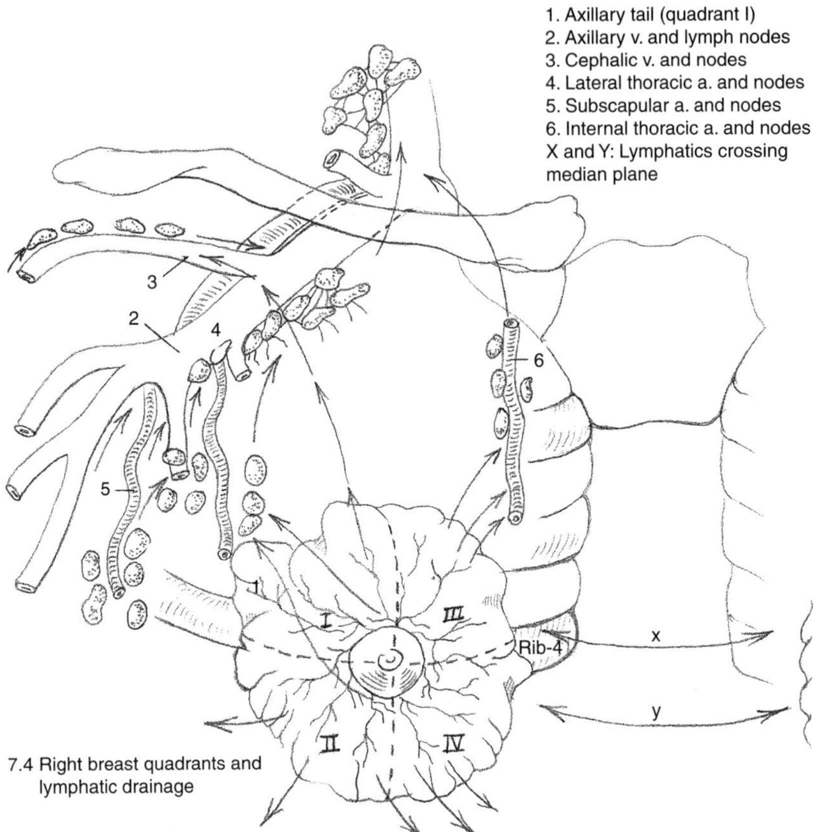

7.4 Right breast quadrants and lymphatic drainage

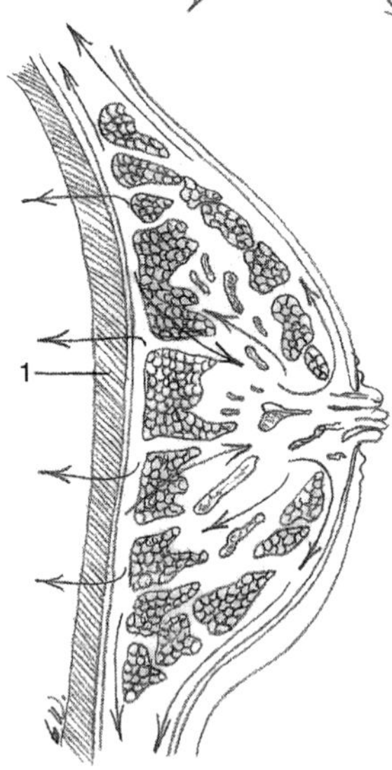

Arrows indicate direction of flow, including that dispersing along Cooper's ligaments.
1. Pectoralis major m. and fascia

7.5 Directions of lymph flow from breast

Figure 7.4: The Human Breast
Figure 7.5: Direction of lymph flow

Chapter 8
THE PERIPHERAL NERVOUS SYSTEM

The mammalian nervous system is divisible into the central nervous system (CNS) and the peripheral nervous system (PNS). The CNS, consisting of brain and spinal cord, is linked to all parts of the body by the PNS, which transmits signals to and from the CNS.

The human PNS is composed of 43 pairs of craniospinal nerves: 31 pairs of spinal nerves that issue in orderly sequence from the spinal cord, and 12 pairs of cranial nerves that emerge from the base of the brain. All branch and arborize profusely as they distribute to the tissues of the body.

Spinal Nerves

The eight pairs of spinal nerves in the neck region are classified as **cervical nerves** (C1–8), the next 12 pairs as **thoracic nerves** (T1–12). There follow in order five pairs of **lumbar nerves** (L1–5), and five pairs of **sacral nerves** (S1–5). The **coccygeal nerves** (Cc, one or more pairs) are the last to come off the spinal cord. All spinal nerves exit the vertebral canal below the vertebra with which they are numerically identified, *except* that the first seven pairs of cervical nerves leave above their matching vertebrae; thus, spinal nerves C7 pass above, and nerves C8 below, vertebra C7. In the thoracic region, spinal nerves T1–12 are associated with same-numbered ribs and intercostal spaces.

All spinal nerves are "mixed"; *i.e.*, they carry both sensory or afferent (*ad-ferro*, I carry towards) and motor or efferent (*ex-ferro*, I carry away) fibers. Sensory nerve fibers pass into the spinal cord as **posterior** (dorsal) **roots** (Fig. 8.1). Their cell bodies lie outside the CNS and collectively make up the **spinal** (dorsal root) **ganglia**. Motor fibers, carrying impulses that evoke action, arise from cells located in anterior (ventral) grey horns of the spinal cord and leave the cord as **anterior** (ventral) **roots**. Anterior and posterior roots unite in the area of the intervertebral foramen to form spinal nerves.

Each spinal nerve divides almost at once into a large **anterior ramus** and a smaller **posterior ramus**. Both rami contain sensory and motor fibers. The anterior rami of several sequentially ordered spinal nerves, and especially those distributing to limbs, join to form plexuses. Peripheral nerves arising from a plexus then usually carry fibers from two or more spinal cord segments; *i.e.*, they are multisegmental nerves, innervating a patch of skin or powering a muscle that otherwise might have to be supplied by several individual segmental nerves. The phrenic nerve (C3-4-5), supplying the diaphragm, is an example of a multisegmental nerve.

Posterior Rami of Spinal Nerves

Each posterior ramus divides into medial and lateral branches, both of which supply, segmentally, the deep (postvertebral) muscles of the back between skull and sacrum. Posterior rami *never* innervate the superficial muscles of the back connecting axial and appendicular skeletons, such as levator scapulae or serratus anterior; nor do they distribute to intercostal muscles, or skin and muscle of limbs—all of which are supplied by anterior rami.

Both medial and lateral branches of posterior rami terminate as posterior cutaneous nerves, which distribute as follows: **Medial branches** innervate skin and fascia of the upper back, nape of neck, and posterior scalp (and, being medial branch fibers, these posterior cutaneous nerves emerge on the surface close to the midline). **Lateral branches** of posterior rami innervate skin and fascia of the lower back (and, being lateral branch fibers, these posterior cutaneous nerves appear on the surface at some distance from the midline).

Exceptions to the foregoing description apply to the posterior rami of the suboccipital nerves (C1), which neither divide, nor have cutaneous branches; and to posterior rami of C7–8 and L4–5, which supply muscle only.

Anterior Rami of Spinal Nerves

All skin and muscle not supplied by posterior rami is innervated by anterior rami and their branches. They distribute either in segmental fashion, *i.e.*, individual nerves supplying unit lengths of muscle, or through plexuses. Thus, anterior rami C1–3 form the **ansa cervicalis** (Fig 8.2), a nerve plexus whose terminal branches are motor to all infrahyoid muscles *except one*: the thyrohyoid muscle is supplied not via the ansa but directly by C1. The same nerve also innervates geniohyoid, the only muscle *above* the hyoid bone supplied by a spinal nerve. Cutaneous innervation of the neck, front and sides, also is through superficial branches of a nerve plexus—the **cervical plexus**, anterior rami C2–4. (*N.B.* Some authors consider the ansa part of the cervical plexus, whose roots then are C1–4).

The roots—meaning here the segmental outflow—of the **brachial plexus** whose terminal branches distribute to skin and muscle of the upper limb are anterior rami C5-6-7-8 and T1; those of the **lumbosacral plexus**, the nerve supply to the lower limb, anterior rami L4–5 and S1-2-3. Anterior rami between the limb plexuses distribute segmentally, *i.e.*, they do not form plexuses; the anterior rami of thoracic spinal nerves, in essence, are represented by intercostal nerves supplying intercostal muscles as well as the skin along the front and the sides of the trunk. Recall, however, that the skin of the back is innervated, segmentally, by posterior cutaneous nerves of posterior rami.

Cutaneous Domains of Spinal Nerves

The area or patch of skin supplied by a single spinal nerve, whether trunk skin or limb skin, is a **dermatome**.

In the trunk, where spinal nerves T2 through L1 distribute segmentally, dermatomes form overlapping bands that encircle the trunk horizontally between the manubriosternal junction (T2) and pubic symphysis (L1). To help you visualize the "dermatome map" (Fig. 8.3), remember that the dermatome crossing the umbilicus belongs to spinal nerve T10. Note, again, that both rami participate in formatting a trunk dermatome—anterior rami by way of anterior and lateral cutaneous nerves, and posterior rami via a posterior cutaneous nerve.

The dermatomes of limbs also are laid out in orderly sequence, but their arrangement is somewhat obscured by events occurring during limb development. More about that in a subsequent chapter.

Cranial Nerves

It is common practice to classify the 12 pairs of cranial nerves numerically, I through XII, the first pair being the olfactory nerves and the last pair the hypoglossal nerves. An alternative classification, used here, is based on functional criteria: some cranial nerves are purely sensory in function; some are predominantly, though not necessarily exclusively, motor; still others are mixed.

NN. I, II, VIII

The **olfactory** (N. I), **optic** (N. II), and **vestibulocochlear** (N. VIII) **nerves** are sensory to the olfactory epithelium, the retina, and the inner ear, respectively. The terms *stato-acoustic* and *auditory* also aptly describe N. VIII, for the organs of equlibrium and balance (*e.g.*, the semicircular canals) reside in the inner ear along with the spiral organ (of Corti).

NN. III, IV, VI

The **oculomotor** (N. III), **trochlear** (N. IV), and **abducens** (N. VI) **nerves** are predominantly motor to the six extraocular muscles. All but two are innervated by N. III that, in addition, supplies the levator palpebrae superiosis muscle, responsible for lifting the upper eyelid. N. IV is the motor nerve to the superior oblique muscle, and N. VI goes to the lateral rectus muscle, the *abductor* of the eyeball.

N. XI

The **accessory nerve** has two roots. One root, known as the spinal root, is a motor nerve—though some authors classify it as mixed. The spinal root, strictly speaking, is not a cranial nerve, for it contains fibers whose cell bodies are located in anterior grey horns of cervical spinal cord segments. The only muscles it supplies are trapezius and sternocleidomastoid.

The cranial root, also known as the bulbar portion of N. XI, is accessory to N. X, the vagus, and will be discussed with that nerve.

N. XII

The **hypoglossal** (below-the-tongue) **nerve** (N. XII), is the motor supply to all muscles that move the tongue; *i.e.*, those that are built into the organ or intrinsic to it, and those that are extrinsic to it. Having adopted this guiding principle, you need to remember only the exception: *m. palatoglossus*, an extrinsic muscle of the tongue, is innervated not by N. XII, but by the accessory root of N. XI.

All extrinsic tongue muscles, incidentally, are instantly identifiable as such by having *glossus* attached to their name.

Mixed Cranial Nerves

All other cranial nerves, *e.g.*, NN. V, VII, IX, and X, are mixed nerves. Each carries both sensory and motor fibers. Be aware, and this is important, that the cell bodies of sensory fibers carried by cranial nerves, like those of spinal nerves, form ganglia outside the CNS—in this case, outside the brain. And, remember, the cranial sensory ganglia are developmentally and functionally identical to spinal ganglia of spinal nerves. They differ from each other only topographically, the former being inside the skull, the latter being inside the vertebral column.

N. V

The **trigeminal nerve**, the *triplet* nerve (N. V), has ophthalmic, maxillary, and mandibular divisions.

The **ophthalmic division** (V1), is entirely sensory in function. Its branches distribute to the cornea and other non-light-sensitive parts of the eye, and to the skin of the upper eyelid, forehead, and dorsum of the nose (Fig 8.4). Beware! The nerve supplies neither the retina nor the extraocular muscles, nor does its lacrimal branch stimulate the lacrimal gland to produce tears.

The **maxillary division** (V2), also contains only sensory fibers. They supply the mucous membrane of the nasal cavities, including paranasal sinuses, the palate, and gums and teeth of the upper jaw. Other fibers distribute to the skin over the temple, the lower eyelid, the upper cheek and lip.

The **mandibular division** (V3), the largest division, has both sensory and motor fibers; *i.e.*, it is a mixed nerve. Its motor root supplies all muscles of mastication as well as additional muscles developing from the first pharyngeal arch (see chapter 12). Its sensory fibers distribute to the gums and teeth of the lower jaw, the mucous membrane of the anterior two-thirds of the tongue, and to the skin over the mandible, including the lower cheek and lip, and parts of the outer ear.

The cell bodies of division V1 and V2 fibers (strictly sensory), and of the mandibular division's sensory fibers, form the prominent **trigeminal** (semilunar) **ganglion** located in the middle cranial fossa.

N. VII

The **facial nerve** (N. VII), consists of a large motor root in combination with the **intermediate nerve** (nervus intermedius). The motor root innervates the muscles of facial expression, including those moving the scalp, and other muscles developing from the second pharyngeal arch (see chapter 12). Nervus intermedius has two branches of special interest: (1) the **greater petrosal nerve** that carries secretomotor fibers to the lacrimal gland and to the mucus-producing tissues of nose and palate; and (2) the **chorda tympani** whose secretomotor fibers supply both the sublingual and the submandibular salivary glands, and whose sensory fibers transmit taste from the anterior two-thirds of the tongue. Deep sensibility of the face also may be subserved by sensory fibers of N. VII. Their cell bodies, and those of taste fibers, are in the **geniculate** (facial) **ganglion**.

N. IX

Stylopharyngeus is the only muscle supplied by N. IX, the **glossopharyngeal nerve**. Parasympathetic fibers composing the **lesser petrosal nerve**, secretomotor to the parotid salivary gland, travel part way with N. IX (see chapter 14). The sensory branches of N. IX include **carotid sinus nerves** from receptors monitoring blood pressure and fibers from the mucous membrane of the posterior one-third of the tongue and upper pharynx transmitting both taste and general sensations. The cell bodies of sensory fibers reside in the **inferior** (petrosal) **ganglion**.

N. X (X-XI)

The **vagus nerve** (*wanderer* or *vagabond*) has an extensive distribution to tissues of the neck, thorax, and abdomen, which it reaches via numerous named branches and plexuses—among them, the **pharyngeal plexus** in the neck and the **cardiac plexus** in the thorax. The cranial root (bulbar portion) of the accessory nerve (N. XI) joins the vagus below the jugular foramen; it probably is the source of most, if not all, motor fibers leaving the vagus above the level of the cardiac plexus. The common trunk, call it the **vago-accessory complex**, innervates the striated muscle of the palate (including palatoglossus, but not tensor veli palatini, which is supplied by the mandibular division of the trigminal nerve), the pharynx (by way of the pharyngeal plexus), and the larynx. According to some authorities, it also supplies the heart (including Purkinje tissue). Below the level of the cardiac plexus, pure vagus fibers distribute to tissues of the lower thorax and abdomen as far as the junction between middle and distal one-thirds of the transverse colon.

Sensory fibers carried by meningeal, auricular, and other vagal branches or plexuses have their cell bodies in the **inferior** (nodose) **ganglion**.

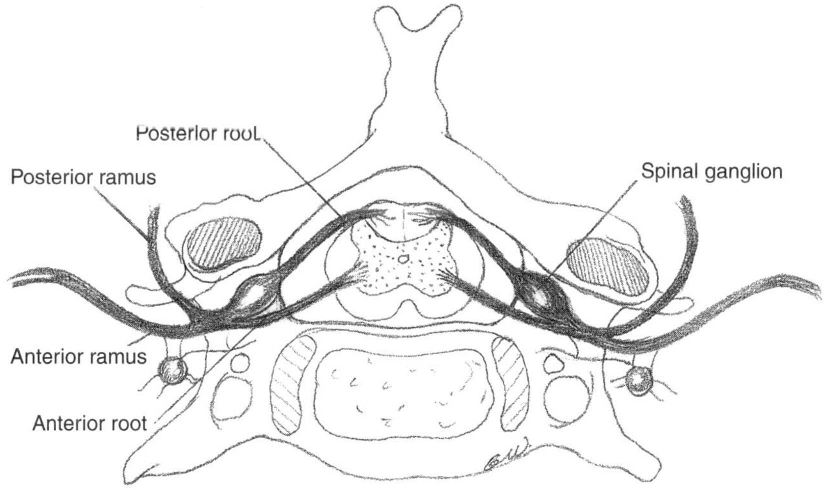

8.1 Anatomy of spinal nerves

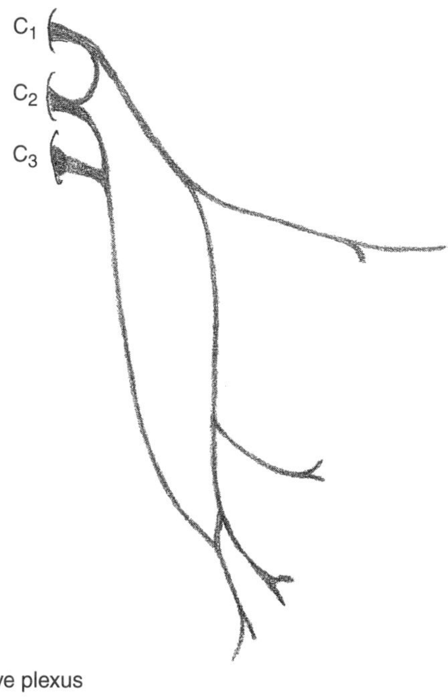

8.2 Anatomy of nerve plexus

Figure 8.1: The anatomy of spinal nerves
Figure 8.2: The anatomy of a nerve plexus

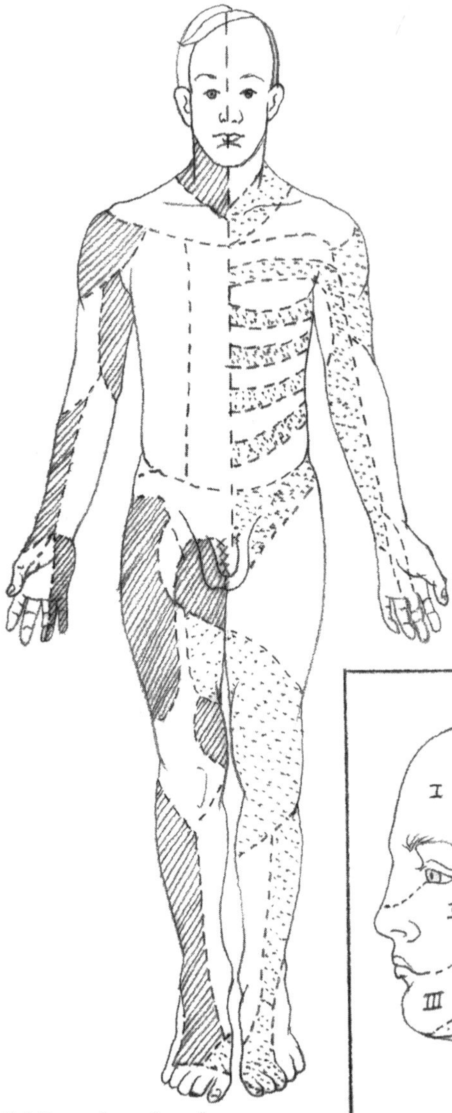

8.3 Dermatome "map"

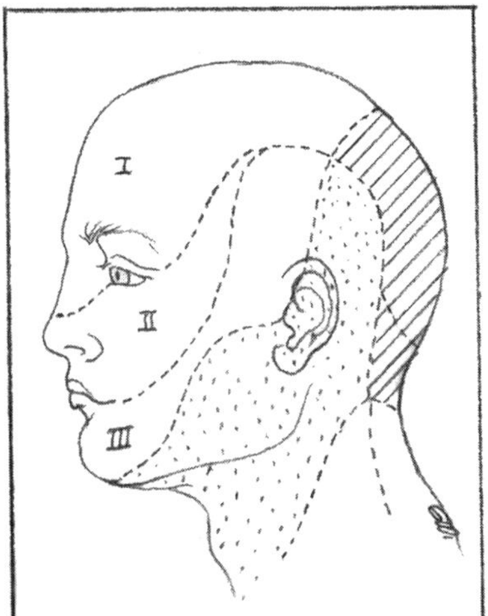

8.4 Cutaneous nerves of head and neck

I. Ophthalmic division of trigeminal nerve
II. Maxillary division of trigeminal nerve
III. Mandibular division of trigeminal nerve
Anterior rami of cervical nerves (stippled)
Posterior rami of cervical nerves (hatched)

Figure 8.3: The dermatome map
Figure 8.4: Cutaneous nerves of head and neck

Chapter 9
THE THORACIC CAGE

The thoracic cage is the protective armor that shields the heart, lungs, and great vessels as well as many abdominal organs. The liver, gall bladder, and spleen, to name just a few, lie under the cover of ribs, because the concavity of the dome-shaped diaphragm is part of the abdominal cavity.

The thorax is flattened in an anteroposterior direction. Its superior aperture is the thoracic inlet; the inferior thoracic aperture or outlet, much wider than the inlet, is bounded laterally by the costal margin.

The Bony Thorax

The bony thorax is part of the axial skeleton. Its components are: the sternum in front, twelve pairs of ribs, and vertebrae T1–12 plus intervertebral discs behind. Each rib is continuous anteriorly with a bar of hyaline cartilage, *i.e.*, a costal cartilage (L., *costa* = a rib); the zone of transition between rib and cartilage is the **costochondral joint** or junction.

Ribs I–VII are *true* ribs, for their costal cartilages attach to the sternum directly. Pairs VIII–XII are labelled *false*, because they connect to the sternum by cartilages other than their own. Pairs XI and XII are classified as *floating* ribs; they end in the muscles of the posterior abdominal wall.

Ribs numbered II–IX are considered typical, in that they each form **costovertebral joints** with bodies of two vertebrae—that of the numerically matching vertebra (N) and that of the vertebra above (N-1). A typical rib, therefore, straddles an intervertebral disc. All but the last two pairs of ribs articulate also with the transverse processes of same-numbered vertebrae, forming **costotransverse joints**.

The lower ribs, in particular, are not horizontally placed; rather, they slope, inferiorly, from back to front and from medial to lateral. As ribs rise during inspiration, transverse and anteroposterior dimensions of the thoracic cavity are increased and intrapulmonary pressure is reduced (volume and pressure vary inversely, Boyle's Law). When intrapulmonary pressure becomes subatmospheric, ambient air inflates the lungs and saturates them with oxygen. The vertical dimension of the thoracic cavity is enlarged with descent of the diaphragm. Because its central tendon is fused to the base of the fibrous pericardium, downward displacement of the diaphragm also causes the heart to elongate, thus adding room medially for lung expansion.

Infants are not efficient *costal* breathers, because their ribs occupy a near-horizontal plane. Contrariwise, pregnant women, whose expanding uterus severely limits the downward movement of the diaphragm, tend to be costal rather than diaphragmatic breathers.

Muscles of Respiration

Eleven pairs of intercostal spaces alternate with 12 pairs of ribs. They are wider in back than in front, because ribs are not parallel. Also note that the anterior ends of intercostal spaces 6 through 11 do not reach the midline, and that the costal margins frame the infrasternal (subcostal) angle. The epigastric region—the pit-of-the-stomach—therefore is left uncovered and unprotected. Here, abdominal muscles and their aponeuroses take the place of bone and cartilage.

Eleven pairs of intercostal muscles, three layers deep, occupy the intercostal spaces (Fig. 9.1). **External intercostals** fill each space between the costotransverse joint behind and the costochondral junction in front; a fairly dense, fibrous sheet, the **external intercostal membrane** attaches the free edge of each muscle anteriorly to the sternum. **Internal intercostal muscles**, deep to external intercostals, run the length of an intercostal space from the sternum in front to the angles of ribs behind; each is succeeded by an **internal intercostal membrane**, which ends opposite the costotransverse joint. A third, incomplete muscle layer lies deep to the internal intercostal muscles and membranes. Its three parts are: (1) **transversus thoracis**, connecting body and xiphoid of sternum to costal cartilages II–VI (*N.B.*, Some texts refer to this part as *sternocostalis* and reserve the term *transversus thoracis* for the deep layer as a whole, as it is analogous to m. *transversus abdominis*, the third and deepest layer of the anterolateral abdominal wall); (2) **innermost intercostal**, placed deep to internal intercostals in the lateral thoracic wall; and (3) **subcostalis**, crossing lower intercostal spaces in back. The three parts are connected by **endothoracic fascia**; it lines the thoracic wall internally and rests on parietal pleura from which it can be separated only with difficulty.

There is no conclusive electromyographic evidence that external and internal intercostal muscles have opposing functions in the human, even though their fibers pass in opposite directions. Both sets appear to be active only during inspiration, when they increase the interior capacity of the thoracic cavity by rotating ribs, especially pairs III–X, upwards and outwards ("pump-" and "bucket-handle" movements). The first two and the last two pairs of ribs move little in quiet inspiration; they respond mainly to the action of accessory muscles of respiration. Expiration is primarily a passive process, ribs returning to positions of rest owing to the effects of gravity and elastic recoil, including the unflexing or unwinding of costal cartilages which become twisted during inspiration. Healthy intercostal muscles also maintain the structural integrity of the thoracic wall, which is alternatingly exposed to a rise and fall in intrathoracic pressure. In fact, bulging intercostal spaces during expiration and their collapse during inspiration are diagnostic signs of intercostal muscle weakness or paralysis.

The principal muscle of respiration is the **diaphragm**. Its major developmental source is the septum transversum, a mass of mesoderm situated in the neck of the early embryo. Later in human development, with elongation of the neck and unflexing of the trunk, the septum transversum is displaced to its definitive location. The muscle drags its cervical nerve supply along—which accounts for the long vertical passage of the **phrenic nerve** (C3-4-5), and for the referral of pain to neck and shoulder following diaphragmatic injury.

Any muscle attaching to ribs or sternum, such as serratus anterior, scalenus anterior, even sternocleidomastoid, qualifies as an accessory muscle of respiration. Pectoralis major, too, attaching to ribs and sternum, helps expand the thoracic cage—especially when the muscle's humeral end is fixed or steadied; thus, individuals recovering from strenuous exercise typically rest with hands on hips or knees so as to create a stable base from which the sternocostal attachment of pectoralis major can act to raise the ribs.

The Neurovascular Anatomy of Intercostal Spaces

Eleven pairs of intercostal vessels and nerves access the body wall. Each bundle seeks the neurovascular plane of an intercostal space, the interval between the middle and third layers of intercostal muscles. The **neurovascular unit**—intercostal vein, artery, and nerve, in that order, V-A-N, top to bottom—lies under cover of the rib forming the upper boundary of an intercostal space (Fig. 9.2). The undersurface of rib XII is grooved by the subcostal vessels and nerve.

Each intercostal space has anterior and posterior intercostal arteries, which anastomose in the neurovascular plane. **Posterior intercostal arteries** 3–11, and **subcostal arteries**, are segmental branches of the thoracic aorta. The upper two interspaces are supplied by branches of the **supreme** (highest) **intercostal artery**, a terminal division of the **costocervical trunk**, itself a branch of the subclavian artery. All give off **dorsal branches** to muscle and skin of the back, **lateral cutaneous branches** to muscles and skin of the lateral body wall, and **collateral branches** to intercostal muscles, parietal pleura, and periosteum. **Anterior intercostal arteries** 1–6 are segmental branches of the internal thoracic artery (Fig. 9.3). Below the level of the xiphisternal joint, opposite interspaces 7 through 9, anterior intercostal arteries. come off the **musculophrenic artery**, a terminal branch of the internal thoracic artery. Both internal thoracic and musculophrenic arteries give rise to **perforating branches**, which pierce overlying muscles and membrane to distribute cutaneously. There are no anterior intercostal vessels in spaces 10 or 11.

The arrangement of intercostal veins is similar to that of intercostal arteries: there are eleven pairs of posterior intercostal veins, one pair of subcostal veins, and nine pairs of anterior intercostal veins.

Anterior intercostal veins drain blood of the upper six interspaces into internal thoracic veins; those of interspaces 7 through 9 terminate in musculophrenic veins. **Posterior intercostal veins**, with some exceptions, and **subcostal veins** end in the **azygos system** composed of the **azygos vein** on the right and **superior** (accessory) and **inferior hemi-azygos veins** on the left (Fig. 9.4). Note these exceptions: (1) venous blood from the first intercostal spaces is drained by **supreme** (highest) **intercostal veins**, which terminate in ipsilateral vertebral or brachiocephalic veins; and (2) venous blood of intercostal spaces 2 and 3 is collected by **superior intercostal veins**. The right superior intercostal vein ends in the arch of the azygos vein, its partner on the left in the left brachiocephalic vein close to the median plane. Note also the lack of symmetry between supreme intercostal arteries supplying intercostal spaces 2 and 3 and supreme intercostal veins only draining interspace 1.

The azygos sytem is a major venous pathway for returning blood from the body wall. The azygos vein passing to the right of the vertebral column connects inferior and superior venae cavae. The two hemi-azygos veins terminate in the azygos vein after crossing the median plane on or about the level of the xiphisternal joint.

The eleven pairs of **intercostal nerves**, and the pair of **subcostal nerves**, represent anterior rami of spinal nerves T1–12, as previously discussed; *i.e.*, they are mixed nerves distributing segmentally. They accompany intercostal vessels and terminate in front as **anterior cutaneous branches**. Their **lateral cutaneous** and **collateral branches** travel with corresponding branches and tributaries of intercostal vessels. **Posterior cutaneous nerves**, however, are branches not of intercostal nerves but of posterior rami T1–12! Odd as it may sound, posterior intercostal vessels run with anterior rami of spinal nerves.

Also, recall that the anterior ends of intercostal spaces 6–11 fail to reach the midline and that abdominal muscles take their place in the epigastric region. Beyond the costal margin, therefore, intercostal nerves 6–11 and subcostal nerves supply muscle and skin of the anterolateral abdominal wall.

Finally, a needle introduced into an intercostal space for the purpose of collecting fluid from the thorax (thoracentesis) must avoid the upper border of the space, where it may damage an intercostal nerve or perforate an intercostal vessel.

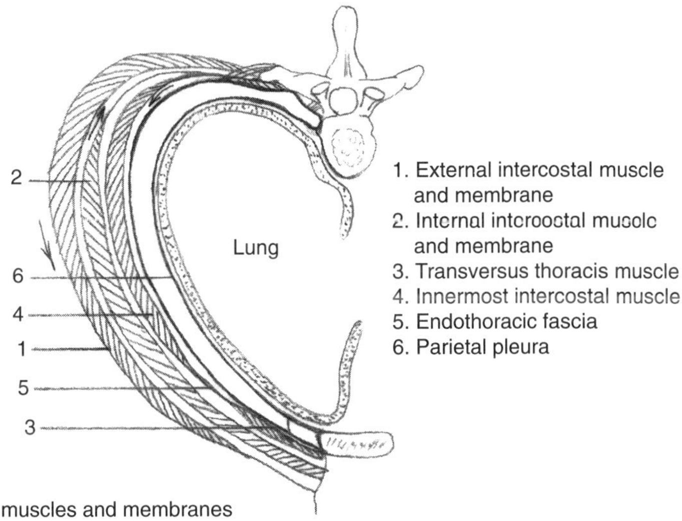

9.1 Intercostal muscles and membranes

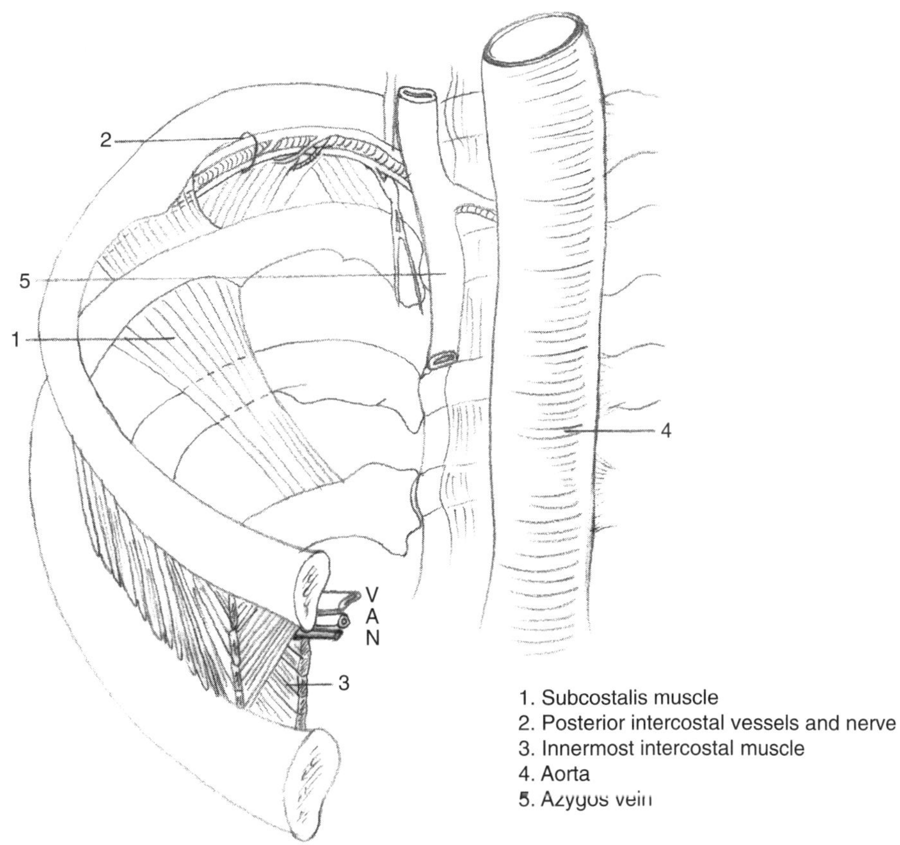

9.2 Anatomy of intercostal spaces

Figure 9.1: Intercostal muscles and membranes, pleural cavity and lung
Figure 9.2: The anatomy of intercostal spaces

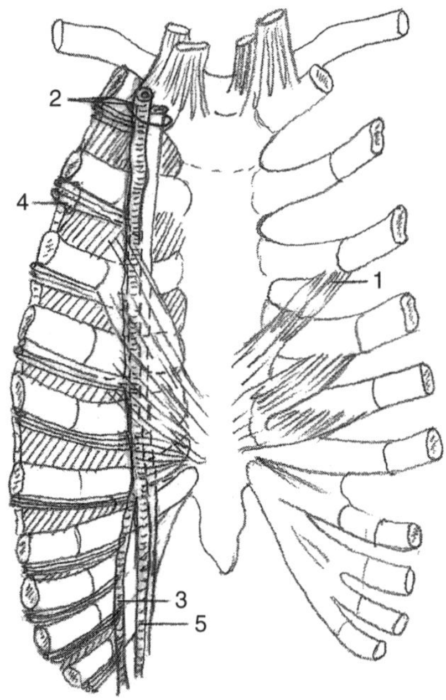

1. Transversus thoracis muscle
2. Internal thoracic vessels
3. Musculophrenic artery
4. Anterior intercostal vessels
5. Superior epigastric artery

9.3 Anterior thoracic wall (posterior view)

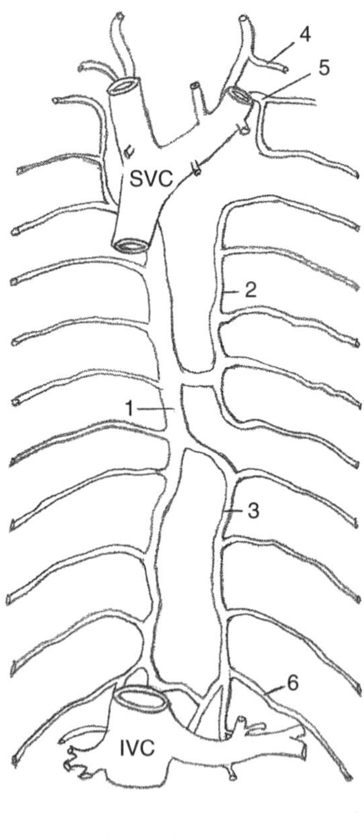

1. Azygos vein
2. Superior hemi-azygos vein
3. Inferior hemi-azygos vein
4. Supreme intercostal vein
5. Superior intercostal vein
6. Subcostal vein

9.4 Azygos system

Figure 9.3: Posterior view of anterior thoracic wall
Figure 9.4: The azygos system

Chapter 10
FORMATION OF HEART AND GREAT VESSELS

The fully-developed human heart is a four-chambered muscular pump. Its chambers are lined with a single row of flat (endothelial) cells, the **endocardium**, in contact with blood. Next to it is cardiac muscle, or **myocardium**, the thickest part of the heart wall. The external layer is the **epicardium**; or, equally correct, the visceral pericardium which, beyond the heart wall, is continuous with parietal pericardium in the same way that visceral pleura covering the lungs is continuous with parietal pleura at the root of the lung (see Fig. 9.1, p. 5.9). The **pericardial cavity** between visceral and parietal layers of serous pericardium is a closed sac containing a small amount of low-viscocity fluid. The parietal pericardium is fused externally with the fibrous pericardium, a protective rather than a serous membrane.

Early Heart Stages

The development of the heart, the establishment of the intraembryonic celom, and folding of the embryo are closely related events. All are initiated towards the end of the third week of human life, when mesoderm has spread to the edges of the elongating presomite embryo and infiltrated the area between the transverse septum and prechordal plate at the rostral end (see Fig. 3.5, p. 20). Spaces developing in lateral plate mesoderm and in front of the prechordal plate coalesce to form the **intraembryonic celom**. Its roof, in contact with ectoderm, is somatopleuric mesoderm; its floor, over yolk-sac endoderm, splanchnopleuric mesoderm (Fig. 10.1).

When viewed from above (Fig. 10.2), the empty intraembryonic celom has the shape of an inverted letter U. The limbs of the inverted U hug the edges of the pear-shaped embryo. Their lateral walls break down as the third week comes to an end (Fig. 10.3); the loop of the U-shaped celom, however, located between septum transversum and prechordal plate, the so-called cardiogenic area, remains a closed tube. Its central section is the future pericardial cavity, and the parts lateral to it become right and left pleural cavities.

The heart and great vessels develop from mesoderm of the cardiogenic area, between the intraembryonic celom and the roof of the yolk sac. Here, mesenchymal cells cluster to form *blood islands*. They grow in length, then link up to form two parallel cardiogenic cords, which canalize to become vessels. Their walls consist of a single, uninterrupted layer of cells—the future endocardium of the heart and endothelium of blood vessels. Other mesenchymal cells condense around the primitive heart tubes as the epimyocardial "mantle," from which will develop myo- and epicardia. The paired primordia move towards the midline, meet, and fuse ventral to the transversely-oriented pericardial cavity (Fig. 10.4).

By gestation-age three weeks the embryo has overgrown the yolk sac and begun to fold in a ventral direction (Fig. 10.5). At the head end, headfold formation shifts the prechordal plate, the future oropharyngeal membrane, to the embryo's undersurface; and, at the opposite end, tailfold formation causes ventral displacement of both the cloacal membrane and body stalk. Headfold formation also causes relocation of the cardiogenic area. In a matter of days, the polarity of the septum transversum at the rostral edge is reversed almost 180°; it now points towards the body stalk, and the pericardial cavity—at one time dorsal to the developing heart—now lies ventral to it. The membranes lining the cavity also have become inverted in the process, so that splanchnopleuric mesoderm, including the heart, now forms the roof of the pericardial cavity and somatopleuric mesoderm its sides and floor. And, when the pericardial cavity expands in a dorsal direction, a double-layered membrane, the **dorsal mesocardium**, suspends the heart tube from the developing foregut (Fig. 10.6).

Once the duplicate character of the heart is lost, its primitive chambers are delineated (Fig. 10.7). They are, front to back: **truncus arteriosus, bulbus cordis, primitive ventricle, primitive atrium,** and **sinus venosus**. At the caudal end, vitelline, umbilical, and common cardinal veins enter the sinus venosus.

The heart is the earliest organ to become functional. It begins to pulsate and to move the fluid within it on or about day 21 of human development. The heart also is the first embryonic structure to lose bilateral symmetry. It bulges ventrally and loops to the right at the same time that it rotates about its longitudinal axis. The primitive atrium, at first caudal to the bulbus cordis now comes to lie dorsal to it (Fig. 10.8). The dorsal mesocardium, stretched by the elongating and flexing heart breaks down in the center. The perforation persists as the **transverse sinus** of pericardium, situated ultimately between the roots of the great arteries developing from the truncus arteriosus and the veins that will enter and contribute to both atria. Clearly, gestation-age three weeks is a critical stage in human ontogeny; we shall revisit it repeatedly in connection with the developmental history of other organ systems.

Septation of Heart Chambers

Internal organization of the heart into four chambers starts at the beginning of the fifth week of embryonic life and continues for about two weeks. It is foreshadowed, when **endocardial cushion tissue**, an acellular subendocardial substance (cardiac jelly), grows into the heart's interior. First to appear, projecting from opposite walls of the atrioventricular junction, are ventral and dorsal **atrioventricular** (A-V) **cushions**. Other cushion tissue, **septum primum** and **secundum**, in succession, descends obliquely from the upper margin of the primitive atrium. Right and left atria result when septal downgrowths fuse with ventral and dorsal A-V cushions (Fig. 10.9 a–c). Blood flow between these chambers is not abolished, however, until birth; an oblique passage in the septal wall, the **foramen ovale**, serves as a shunt for oxygen-rich blood from the right atrium to the left side of the heart. The **fossa ovalis** in the postnatal heart marks the original site of the foramen ovale. The primitive ventricle is divided by endocardal cushion tissue growing in from below, though the right and left ventricles communicate for a time by way of the **interventricular foramen**. The **truncobulbar** segment of the heart, likewise, is partitioned by the **aorticopulmonary** (*spiral*) septum, separating future **pulmonary trunk** from the **ascending aorta** (Fig. 10.10). Observe that the lower edge of the septum helps seal the interventricular foramen, towards the end of the seventh week, by fusing with the A-V cushions and the tip of the interventricular septum. And, once the interventricular foramen is closed, the outflow path from the left ventricle, known in descriptive anatomy as the **aortic vestibule** or **recess**, is isolated from the outflow path of the right ventricle, the **conus anteriosus** or **infundibulum**.

Development of Great Arteries

Development of the cardiovascular system begins, you'll recall, with the formation of two parallel vessels in the splanchnopleuric mesoderm of the cardiogenic area. They fuse in the head region to establish the single heart tube, but remain independent of each other as they grow tailward. When the cardiogenic area is turned over, the unfused proximal parts of the two vessels along with the primitive heart are brought into a ventral relationship with the future foregut (Fig. 10.11). And, when headfold formation is complete, the originally straight vessels resemble fishing hooks. The long shank of each hook is a **dorsal aorta**; the two dorsal aortae will fuse distally to form the descending aorta. The short, pointed end of each hook is a **ventral aorta**; the two ventral aortae will fuse just ahead of the truncobulbar segment of the heart to become the **aortic sac**. The loops connecting ipsilateral dorsal and ventral aortae are the first pair of **aortic arch arteries**. Five additional pairs of aortic arch arteries will appear subsequently in series with the first pair. Not all six pairs persist, however, nor do all parts of the ventral and dorsal aortae. Figure 10. 12 depicts the aortic arch system in ventral view and illustrates the fate of aortic arch arteries. Alternative descriptions of aortic arch transformations do exist but they lack the simplicity of the following schemes:

The **arch of the aorta** appropriates for itself (1) the ventral root of the left fourth arch artery (*root* meaning here the segment of the left ventral aorta between arches four and five), (2) all of the left fourth arch artery, and (3) the dorsal roots of left arch arteries four and five (*roots* referring here, again, to segments of left dorsal aorta between arches). The arch of the aorta is continuous distally with the **descending aorta** formed by the left dorsal aorta and, more distally, by fused right and left primitive dorsal aortae. Proximally, the arch of the aorta is the continuation of the **ascending aorta**.

The ventral root of the right fourth arch becomes the **brachiocephalic trunk**. The right fourth arch artery turns into the proximal segment of the right **subclavian vessel**. Its distal parts come from the dorsal roots of right arches four and five, and the right seventh intersegmental artery. The left seventh intersegmental vessel gives rise to the left **subclavian artery**. The ventral roots of third arch arteries transform into **common carotid arteries**, those of the second arch into **external cartotid vessels**. The **internal carotid arteries** are derived from third arch arteries plus the dorsal roots of the first two arches. The sixth aortic arch arteries, except for the left distal segment, become **pulmonary arteries**—continuous with the pulmonary trunk produced by septation of the truncobulbar segment. The distal segment of the left sixth aortic arch artery persists as the **ductus arteriosus**. During fetal life, this large vessel shunts oxygen-rich, placental blood from the right side of the heart into the systemic circulation.

All other parts of the aortic arch system, with some minor exceptions, degenerate and disappear.

Development of Major Veins

At first, a single pulmonary vein enters the dorsal wall of the future left atrium (Figs. 10.9, 10.13); whether it is an outgrowth from the heart or an ingrowth from the developing lungs is not clear. In either event, the single stem receives tributaries from both lungs, as shown. With enlargement of the left atrium, the trunk and tributaries are absorbed into the chamber's wall until the definitive arrangement—openings for right and left pairs of **superior** and **inferior pulmonary veins**—is attained (Fig. 10.9 c). The interior of the fully-formed left atrium reflects its developmental origin: the smooth part of the chamber is the contribution made by primitive pulmonary veins; the trabeculated wall and **auricle** are remains of the original atrium.

Early in the development of the cardiovascular system, the **sinus venosus**, the venous end of the primitive heart, consists of (1) a middle section, in line with the primitive atrium, and (2) right and left horns. Each horn, at that time, receives **vitelline, umbilical,** and **common cardinal veins** (Fig. 10.7). The cranial end of the right vitelline vein persists as the terminal segment of the

inferior vena cava. More about the rest of the vitelline system and umbilical veins in subsequent chapters.

The common cardinal veins (ducts of Cuvier) drain anterior and posterior cardinal veins. **Anterior cardinal veins** return blood from head, neck, and upper limb. **Posterior cardinal veins** persist only at their extreme cranial and caudal ends and have more limited fields of drainage: the cranial end of the right posterior cardinal vein remains as the **arch of the azygos vein**, that of the left as the **left superior intercostal vein**. The extreme caudal ends of the posterior cardinals transform into **common** and **internal iliac veins**.

The development of the internal jugular, brachiocephalic, and subclavian veins takes this form (Fig. 10.14): both **subclavian veins** develop independent of the cardinal system from seventh intersegmental veins and attach secondarily to anterior cardinal veins. Their entrance marks the division of anterior cardinals veins into **internal jugular** and **brachiocephalic veins.** Most of the left brachiocephalic vein, however, is newly derived, in large part by fusion of small neck veins draining midline organs such as **thyroid** and **thymus glands**. Where right and left brachiocephalic veins meet, the **superior vena cava** (SVC) begins. As you can tell, the right anterior cardinal furnishes only the cranial segment of the SVC; the right common cardinal vein forms the rest. It should not surprise you, therefore, to find the arch of the azygos vein enter the SVC near its midsection.

Fate of the Sinus Venosus

Differential growth rates during the convoluted stage of heart development caused the sinus venosus to shift to the right while rotating some 80 degrees (Figs. 10.9, 10.14). It now is in line with the future right atrium and projects into it. The margins of the sinus venosus at the entrance into the right atrium become the right and left **sinal valves**. The right horn of the sinus venosus grows much larger than the left horn; and, as the right atrium continues to expand, the sinus venosus is gradually absorbed and becomes part of the right atrium. Both venae cavae now open directly into the right atrium, as does the **coronary sinus** whose antecedents are the left horn of the sinus venosus and the adjacent section of the left common cardinal vein.

The left sinal valve disappears, as it is incorporated into the interatrial septum. The right sinal valve persists as a landmark inside the right atrium, the **crista terminalis**, which separates the contributions made to that chamber by sinus venosus and primitive atrium. The trabeculated part of the definitive right atrium, like that of the left atrium, including auricle, is original primitive atrium whereas the smooth portion of the atrial wall is original sinus venosus.

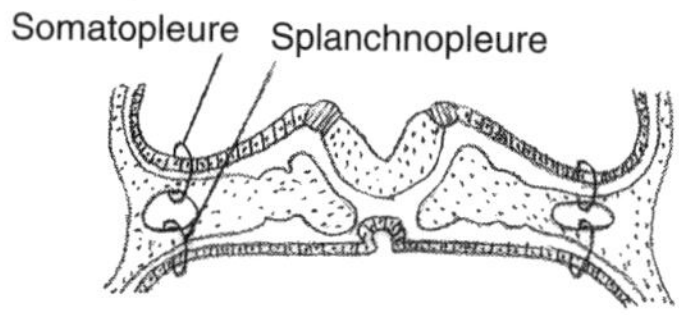

10.1 Developing celom

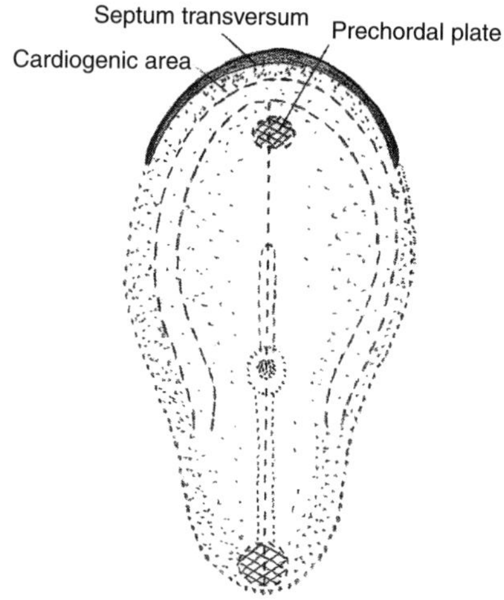

10.2 Embryonic celom (dorsal view)

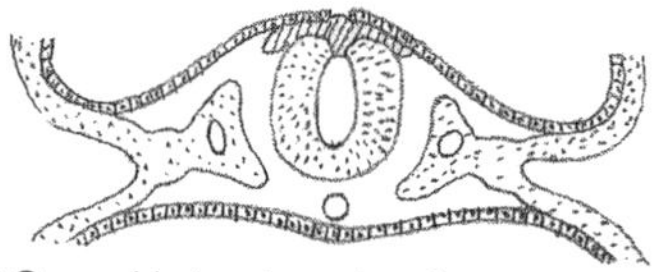

10.3 Open-sided embryonic celom

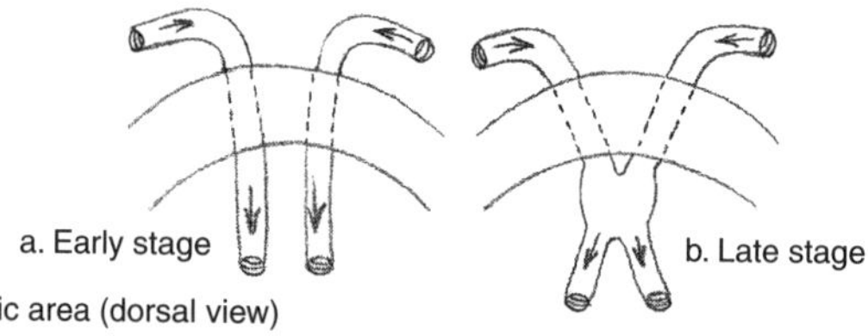

10.4 Cardiogenic area (dorsal view)

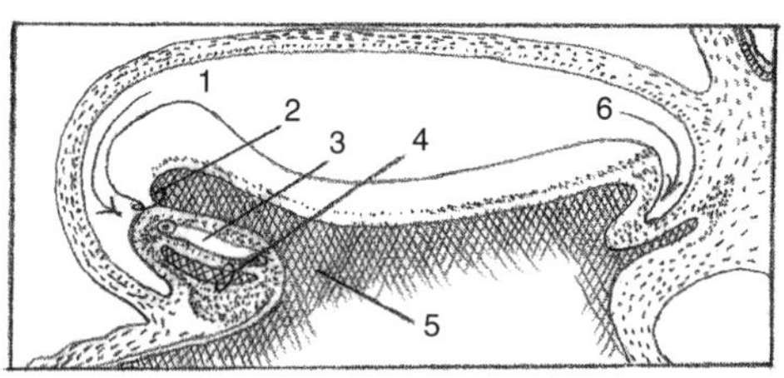

10.5 Folding embryo (longitudinal section)

1. Headfold
2. Prechordal plate
3. Heart tube
4. Pericardial cavity
5. Foregut
6. Tailfold

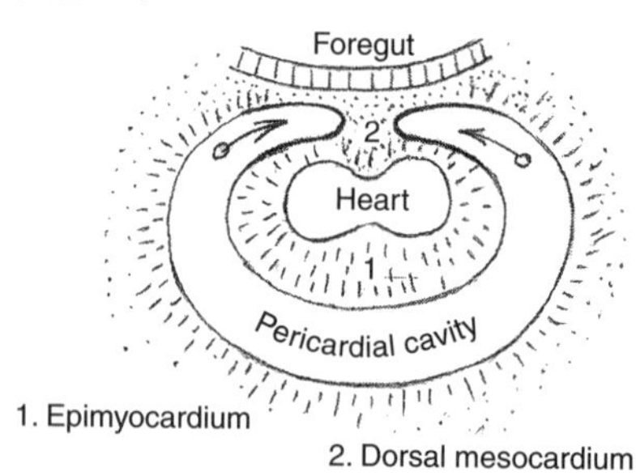

10.6 Heart tube and pericardium (cross-section)

Figure 10.1: The development of the embryonic celom
Figure 10.2: Dorsal view of embryonic celom
Figure 10.3: The open-sided embryonic celom
Figure 10.4: Dorsal view of cardiogenic area
Figure 10.5: Longitudinal section through the folding embryo
Figure 10.6: Cross-section of heart tube and pericardium

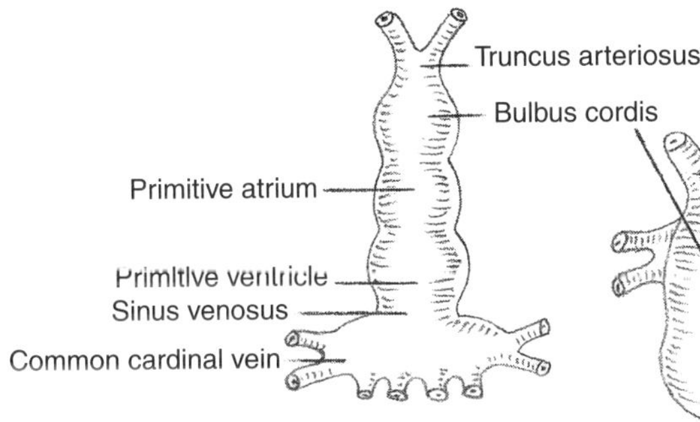

10.7 Primitive heart chambers

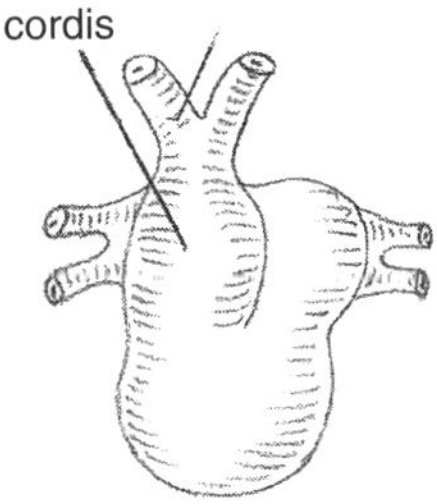

10.8 Convoluted stage

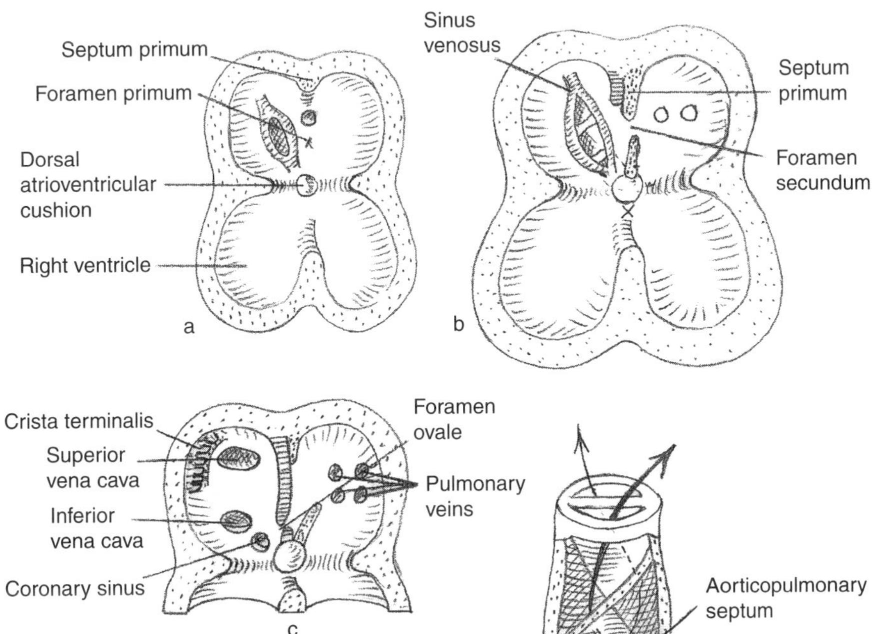

10.9 a-c Stage in heart development

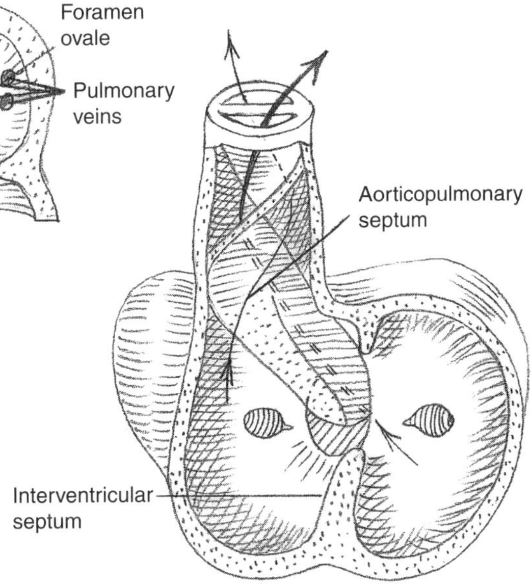

10.10 Aorticopulmonary septum

Figure 10.7: The primitive heart chambers
Figure 10.8: Convoluted stage
Figure 10.9: a + b + c Stages in heart development
Figure 10.10: Aorticopulmonary septum

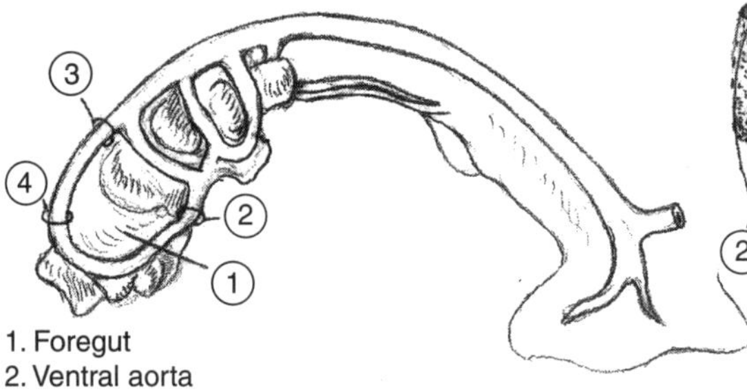

1. Foregut
2. Ventral aorta
3. Dorsal aorta
4. First aortic arch

10.11 Aortic arch arteries (lateral view)

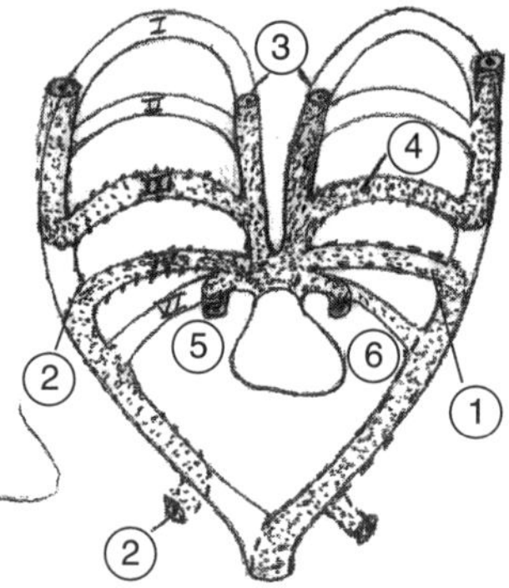

1. Arch of the aorta
2. Right subclavian a.
3. External carotid aa.
4. Internal carotid a.
5. Pulmonary a.
6. Ductus arteriosus

10.12 Fate of aortic arch arteries

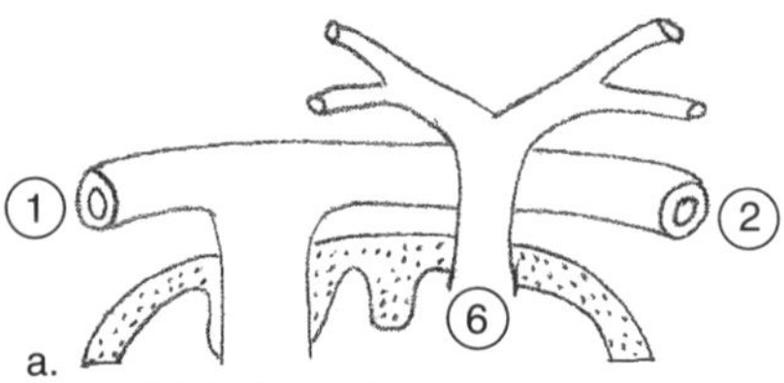

1. Right horn of sinus venosus
2. Left horn of sinus venosus
3. Septum primum
4. Septum secundum
5. Coronar sinus
6. Pulmonary vein

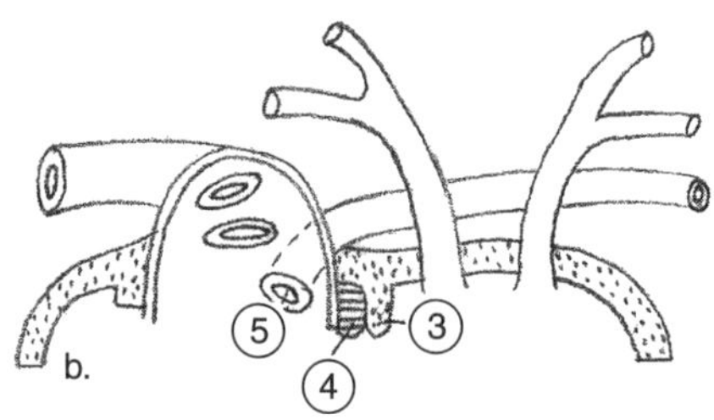

10.13 a-b Rear wall of primitive atrium at level of sinus venosus (horizontal section):
a. Before, b. After absorption of sinus venosus and pulmonary veins

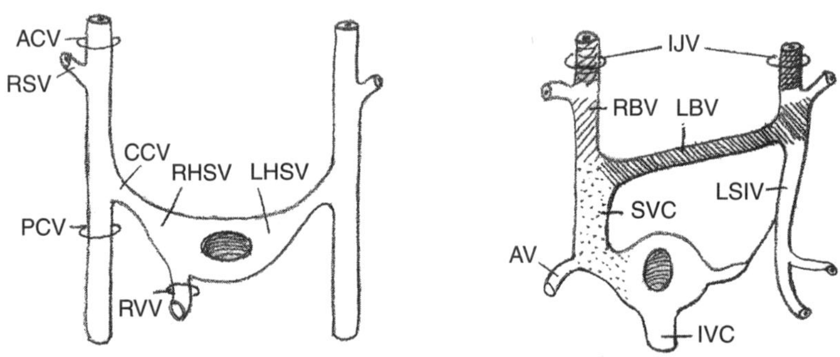

a. Early stage

b. Late stage

10.14 a-b Development of great veins (see text for detail)

Figure 10.11: Lateral view of aortic arch arteries
Figure 10.12: Fate of aortic arch arteries
Figure 10.13: a + b Contribution made by sinus venosus and pulmonary veins to right and left atria
Figure 10.14: a + b Development of great veins

Chapter 11
THE MEDIASTINUM

The complex arrangement of the thoracic viscera and their important topographic relationships are perhaps best revealed, understood, and retained when expressed in terms of some relatively stable or fixed skeletal reference point, *e.g.*, a numbered vertebra. If the reference object is visible and/or palpable on the body's surface, so much the better. The **sternal angle** (of Louis), a slight bony ridge or depression on the upper chest wall, corresponding to the manubriosternal joint, is such a landmark. It is useful, because it indicates on the body's surface the anterior projection of an imaginary horizontal plane at which many important mediastinal structures begin, end, or change direction.

Remember, however, that the location of the manubriosternal joint is variable in life; that it rises, for instance, in relation to the vertebral column during deep inspiration and falls during expiration. Also, it is higher when the body is horizontal rather than upright.

Be aware also that thoracic viscera change position and shape constantly. Their stated locations, even when expressed in terms of some skeletal marker, are only approximations; *e.g.*, the bifurcation of the trachea descends in full inspiration below the level of vertebra T6 when the body is in the upright position, but is approximately one vertebral level higher in the recumbent individual.

The **mediastinum** is the space and substance between the lungs; or, more specifically, between right and left mediastinal pleurae. Its limits above and below are the superior and inferior thoracic apertures, respectively. It is divisible into superior and inferior mediastina, conceptually separated from each other by an imaginary horizontal plane passing from the sternal angle in front to the thoracic spine behind (Fig 11.1). The inferior mediastinum includes anterior, middle, and posterior mediastina. The **middle** mediastinum contains the heart and the roots of the great vessels; the **anterior** mediastinum is situated between the body of the sternum and the heart, the **posterior** mediastinum between the heart and the vertebral column.

A helpful exercise in reconstructing mediastinal contents is to sketch, layer by layer, starting with the vertebral column behind, the major thoracic organs, vessels, and nerves in frontal view. Drawing the cross-sectional anatomy of the thorax at succeeding vertebral levels is equally effective in gaining a three-dimensional image of topographic relationships (Fig. 11.2 a–d). This chapter will show that the sternal angle is a useful tool in the modelling of thoracic contents.

The Transangular Plane

As earlier noted, the sternal angle marks the position of the manubriosternal joint—the fibrocartilaginous union between body and manubrium of the sternum. It also marks the attachments of the second pair of costal cartilages and, therefore, of the second pair of ribs; hence, ribs above and below the second pair can readily be located by palpation and identified by number. The imaginary horizontal plane we are discussing, the **transangular plane**, stretches from the sternal angle in front to the vertebral column behind. It separates the superior mediastinum above from the inferior mediastinum below. In the cadaver, its backward projection intersects the thoracic spine approximately at the level of the disc between vertebrae T4 and T5; it is at least one vertebral level lower in the upright, living individual.

Here, then, is a partial list of structures that begin, end, or change direction close to the transangular plane. We start with the **esophagus** which, in anterior view, overlies the vertebral column Fig 11.3). The esophagus is about 26 cm (10 in.) long. It begins in the neck opposite vertebra C6 as a continuation of the pharynx and lies somewhat to left of midline as it enters the thorax. Thereafter, the organ curves to the right and attains a midline position at the level of the sternal angle. Before passing through the diaphragm opposite vertebra T10 and entering the abdomen, the esophagus again moves to the left.

The **trachea** is roughly one-half the length of the esophagus (*i.e.*, 10–12.5 cm or 4–5 in) upon which it is superimposed in Fig. 11.3. The trachea, like the esophagus, begins at vertebral level C6. It is the continuation of the larynx and, after a short (5–6.25 cm or 2–2.5 in.) passage through the neck, the organ enters the thorax strictly in the midline. It terminates opposite the sternal angle by dividing into left and right main bronchi. The lower end of the trachea lies in the superior mediastinum, slightly to the right of the midline owing to pressure from the aortic arch passing on the left. Hence, the **left main bronchus** is longer (and narrower) than the right. The angle of bifurcation is approximately 70°. The **right main bronchus** takes a more vertical course than does the left (25° deviation from the median plane on the right vs. 45° on the left); thus, aspirated objects are more apt to lodge in the right lung. The left main bronchus passes obliquely across the esophagus before entering the substance of the left lung and dividing into two secondary **lobar bronchi**. The upper lobar bronchus supplies the upper lobe and lingula, the other distributes to the lower lobe.

The right main bronchus divides *before* it enters the lung. The lobar bronchus going to the upper lobe is above the right pulmonary artery, hence its name "ep(i)arterial." The other one, called "hyp(o)arterial" because of its relationship to the same artery, branches to supply both middle and lower lobes of the right lung. Right and left lobar bronchi give rise to a total of 18–20 **segmental bronchi**, each of which distributes to one of the structural and functional units of the lungs, the bronchopulmonary segments.

The **arch of the aorta** is located in the superior mediastinum; therefore, the arch begins and ends close to the transangular plane. We may conclude also, that the ascending aorta ends, and the descending (thoracic) aorta begins, at that level. Because the aortic arch curves from right to left in an anteroposterior direction, the right main bronchus passes behind the ascending aorta and the left main bronchus in front of the descending aorta.

Three large vessels spring from the arch of the aorta: the brachiocephalic trunk, dividing into right subclavian and right common carotid arteries, the left common carotid artery, and the left subclavian artery (Fig. 11.3). We may enter also their companion veins; *i.e.*, right and left subclavian and internal jugular veins. Subclavian and internal jugular veins of the same side unite to form a brachiocephalic vein. The left, crossing obliquely to the opposite side, combines with the right brachiocephalic vein to become **superior vena cava** (SVC) behind and below the right sternoclavicular joint. Therefore, the SVC begins an inch or so above the sternal angle and is an exception to the doctrine that many important structures begin, end, or change directions at the level of the sternal angle. As I pointed out in chapter 1, remember the exceptions! They greatly simplify life.

Next, the **pulmonary trunk** divides into right and left **pulmonary arteries** near the **transangular plane** (Fig. 11.3). The heart being on the left side, the right pulmonary artery is longer than the left, and it must cross the median plane on its way to the hilum of the right lung. Connecting the left pulmonary artery and the arch of the aorta close by is the **ligamentum arteriosum**, the fibrous remains of the ductus arteriosus, a now obliterated fetal vascular shunt. The **left recurrent laryngeal nerve**, coming off the left vagus nerve as that nerve passes over the arch of the aorta, winds around the ligamentum arteriosum before ascending to its destination—the larynx. In its upward course, the left recurrent laryngeal nerve lies somewhat to the left of the trachea and in front of the esophagus (Fig. 11.2 a–b).

In the neck, right and left vagus nerves are accompanied by right and left common carotid arteries and internal jugular veins inside the densely fibrous carotid sheath. As the two nerves enter the thorax, both are behind the left brachiocephalic vein, and each is in front of the root of a subclavian artery (Fig. 11.3). For reasons discussed a moment ago, the right vagus passes behind, and the left nerve passes in front of the aortic arch. Thereafter, both descend behind the root of a lung and form a plexus on the esophagus; right and left vagi continue as posterior and anterior gastric nerves, respectively, inside the abdomen. Recall that the left recurrent laryngeal nerve loops under the ligamentum arteriosum; the recurrent laryngeal nerve on the opposite side leaves the right vagus some distance above the transangular plane and winds around the root of the right subclavian artery before reentering the neck.

The **arch of the azygos vein**, which ends in the superior vena cava (Figs. 11.2 b, 11.3), makes its turn on or about the transangular plane. The azygos vein is a large vessel collecting blood from the posterior body wall, including intercostal spaces. It begins in the abdomen opposite the renal veins, leaves the abdomen by way of the aortic hiatus, and ascends in the right posterior mediastinum in contact with the vertebral column. The vein reaches the superior vena cava from behind by arching over the root of the right lung. The **right superior intercostal vein**, which receives blood from the second and third right intercostal spaces, drains into the arch of the azygos vein and ends, therefore, at the level of the sternal angle. The azygos vein terminates in the superior vena cava near its midpoint; hence, neither the beginning nor the end of the SVC lies on the transangular plane.

The **thoracic duct** (Figs. 11.2 a–d, 11.3) is the principal vessel of the lymphatic system, returning lymph from most of the body to the systemic venous circulation. It arises by the union of two or more large trunks inside the abdomen, then travels with the azygos vein through the aortic hiatus and posterior mediastinum. Upon reaching the transangular plane, the thoracic duct moves across the midline from right to left and, at the base of the neck, delivers its contents to the left subclavian vein. Lymph originating from parts of the body not drained by the thoracic duct is carried by the right lymphatic duct to the right subclavian vein.

The **apex of the pericardium** extends upwards to the level of the transangular plane, and the transverse sinus of the pericardium is near it. Right and left sternal lines of pleural reflection, being separated from each other at the root of the neck, meet in the midline behind the sternal angle. But, they retain this relationship only briefly; for, opposite the fourth sternocostal joint, the left line of reflection moves away from the median plane. Farther down, in the area of the cardiac notch where no lung tissue intervenes between the right ventricle of the heart and chest wall—the area of complete cardiac dullness—left costal and mediastinal pleurae are apposed to form the costomediastinal recess.

Finally, consider the phrenic nerves. Each enters the thorax between subclavian artery and vein. Thereafter, the right phrenic clings to the right brachiocephalic vein, superior vena cava, and the pericardium over the right atrium (Fig. 11.2 a–d). The left phrenic nerve crosses the vagus at the root of the left common carotid artery and, upon reaching the transangular plane, the nerve veers sharply to the left and descends on the pericardium over the left ventricle. The phrenic nerves, therefore, map out the right and left margins of the heart and middle mediastinum. They are easily distinguished from the vagus nerves by being more anteriorly placed and by passing *in front* of the lung roots (Fig. 11.2 a–d).

The Xiphisternal Plane

An imaginary horizontal plane stretching between the xiphisternal joint in front and vertebrae T8–9 behind is also a useful guide to mediastinal contents and relationships (Fig 11.1). It crosses the top of the dome-shaped diaphragm, the base of the fibrous pericardium, and the inferior border of the heart—including the entrance of the inferior vena cava into the right atrium. The xiphisternal plane is level with the upper border of the liver on the right and the fundus of the stomach on the left. Further, at or near vertebra T8, the two hemi-azygos veins pass from left to right to join the azygos vein (Fig. 11.2 d), and the descending aorta gains a midline position in front of the vertebral column.

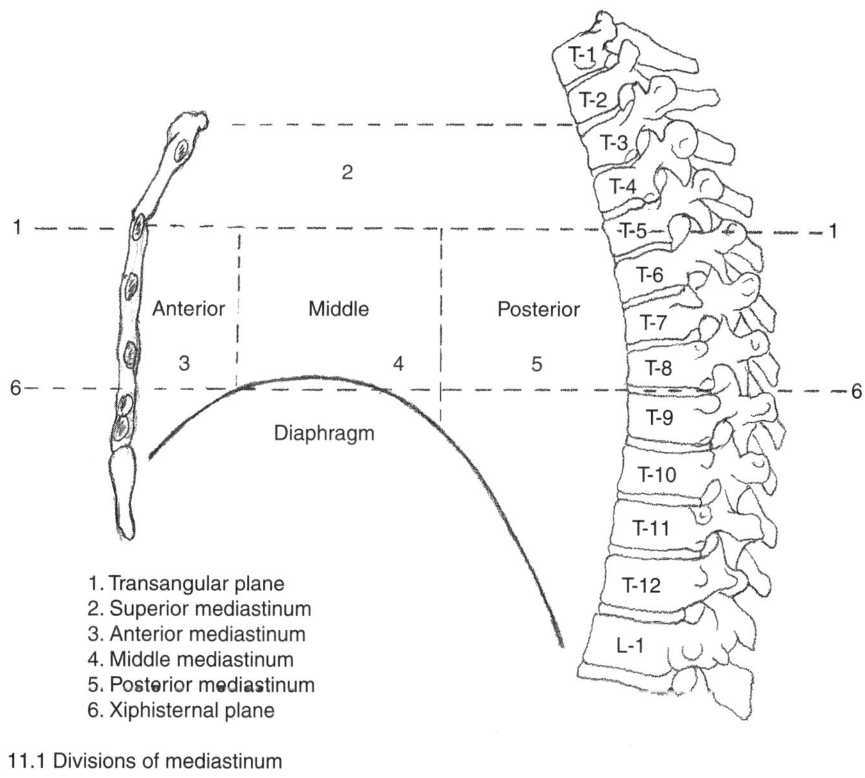

Figure 11.1: Divisions of the mediastinum

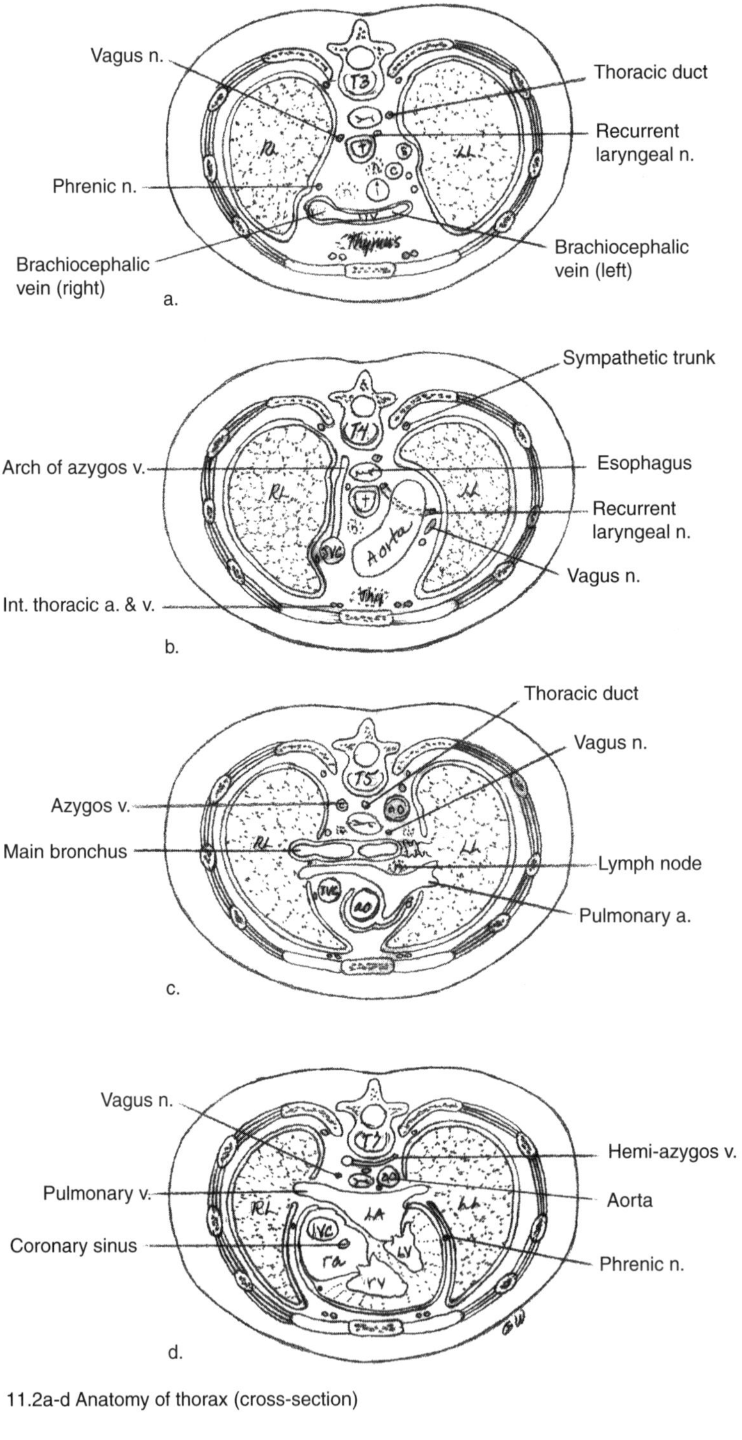

11.2a-d Anatomy of thorax (cross-section)

Figure 11.2: Cross-sectional anatomy of the thorax

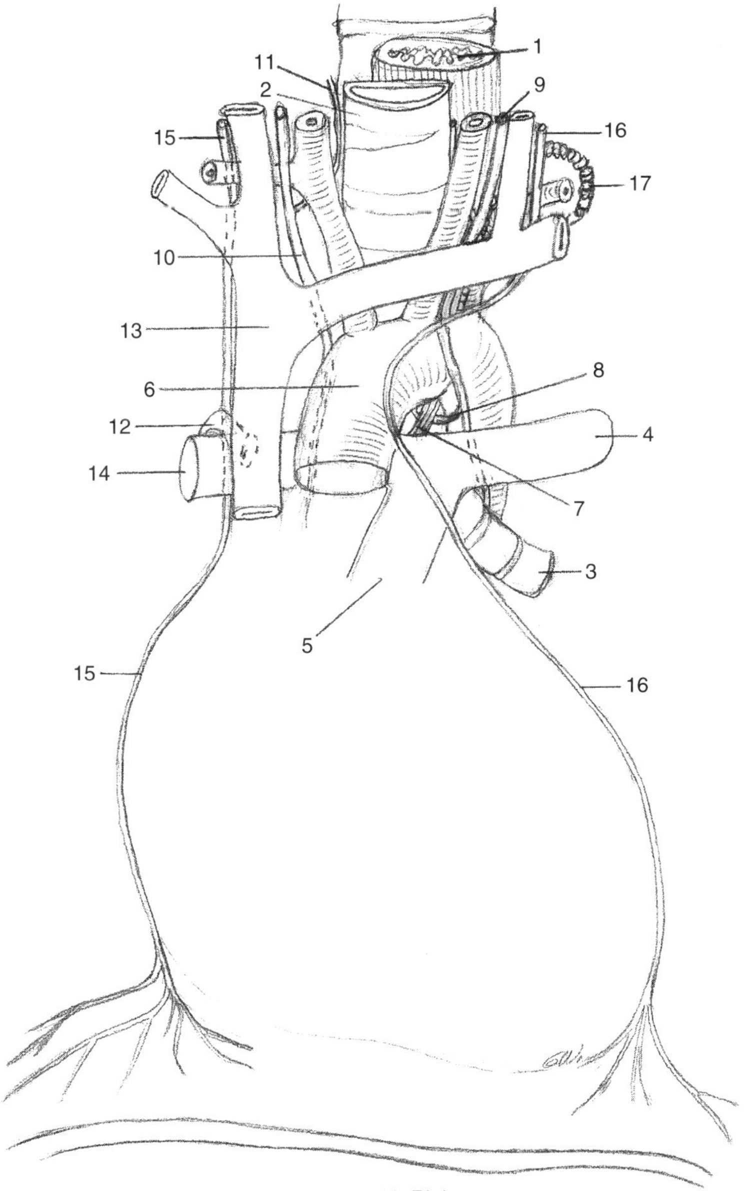

1. Esophagus
2. Trachea
3. Left main bronchus
4. Left pulmonary a.
5. Pulmonary trunk
6. Aortic arch and great vessels
7. Ligamentum arteriosum
8. Left recurrent laryngeal n.
9. Left vagus n.
10. Right vagus n.
11. Right recurrent laryngeal n.
12. Arch of azygos v.
13. Superior vena cava and tributaries
14. Right pulmonary a.
15. Right phrenic n.
16. Left phrenic n.
17. End of thoracic duct

11.3 Topographic relations at transangular plane (T4–5)

Figure 11.3: Topographic relations (transangular plane)

Chapter 12
HEAD AND NECK ANATOMY

The neck is a complicated part of human anatomy, for it is crammed full of muscles, vital organs, vessels and nerves, many of which are in transit between the head and the rest of the body. For convenience of description, and to make the study of this congested region more manageable, the neck may be visualized as a composition of triangles—anterior, posterior, and lateral—each of which can be examined for contents.

An equally effective method for mapping out the contents of the neck is to study the natural subdivisions created by the arrangement of the deep cervical fascia. You will recall that deep fascia serves a useful purpose in topographic anatomy, for it compartmentalizes units of similar structure and/or function. Vertebral levels and imaginary horizontal planes also can be utilized to great advantage in the study of the neck. We shall examine its cross-sectional anatomy at the vertebral level C6 later in this chapter; but first, a brief review of head and neck development.

Head and Neck Development

Cell proliferation in the future head region of the early embryo produces a number of tissue blocks known as *processes* or *elevations*, separated from one another at first by grooves or fissures. Other growth centers create five or six pairs of elevations in the future neck region of the embryo. They partially encircle the neck and form arches, which in water-dwelling animals are associated with gills. The Greek equivalent for gills is *branchia*—hence, the term branchial arches; in human anatomy, **pharyngeal arches** seems a more appropriate designation.

The core substance of the tissue blocks in the embryo's head region is mesenchyme, *i.e.*, primitive, undifferentiated connective tissue. Its major germ layer source is non-somite mesoderm, but endoderm of prechordal plate origin and neural crest ectoderm also contribute to head mesenchyme. Neural crest cells invade the pharyngeal arches early in life and become involved in the development of the facial skeleton and the striated muscle of head and neck.

The first and most rostral pair of pharyngeal arches, also known as **mandibular arches**, appears on or about day 23 of embryonic life—the second pair, the **hyoid arches**, a day or so later. Their appearance during the beginning of the fourth week coincides chronologically with the shift of the prechordal plate to the embryo's ventral surface. It now, as oropharyngeal membrane, forms the floor of the oral pit or stomodeum—the future mouth. When the membrane breaks down, at or about gestation-age four weeks, stomodeum and primitive foregut are confluent.

Each mandibular arch has **mandibular** and **maxillary processes**. As shown in Fig. 12.1, the stomodeum is bounded caudally by the fusing mandibular processes, laterally by the maxillary processes, and cranially by the **frontonasal process** which, in due course, becomes forehead, nose, the middle third of the upper lip, and premaxilla. The paired mandibular processes fuse ventrally in the midline during the sixth week of human development to become the lower jaw, both bony and fleshy; and the paired maxillary processes develop into the upper jaw, including the lateral thirds of the upper lip, and the facial skeleton to the level of the orbits. Furthermore, by fusing internally across the primitive oral cavity and joining the premaxilla, the two maxillary processes create the palate that will separate mouth from nose (Fig. 12.2). Failure of maxillary processes to fuse produces congenital abnormalities.

Pharyngeal Arches

Figure 12.3 a is a dorsal view of the wall and floor of the primitive pharynx after its roof has been separated (Fig. 12.3 b) and removed.

The pharyngeal arches in the embryo's neck region are localized thickenings of mesenchyme covered externally with ectoderm becoming epidermis. Their internal surface is endoderm, forming the ephithelial lining of the foregut—the part of the yolk sac entrapped and internalized during the headfold stage (see Fig. 10.5, p. 66). In the neck, the foregut becomes primitive pharynx.

Pharyngeal arches are separated externally by grooves and internally by furrows or pouches. Their future development, while interesting, is beyond the scope of this chapter. Pharyngeal mesenchyme, however, is important in craniofacial and neck development: it differentiates into bone, cartilage, ligaments, and striated muscle and needs to be considered further.

Embedded in pharyngeal arch mesenchyme is a bar of cartilage, a cranial nerve, and an aortic arch artery. The cartilage bar of the mandibular arch, known as Meckel's cartilage, is not involved in mandible formation, which is intramembranous not endochondral. It does figure, however, in the development of the spine of the sphenoid bone and of the two ear ossicles—the malleus and incus (Fig. 12.4). The sphenomandibular ligament is believed to be a remnant of Meckel's cartilage.

The resident motor nerve of the mandibular arch is the mandibular division of N. V. It will innervate all striated muscles forming in first arch mesenchyme; namely, the muscles of mastication—**temporalis, masseter**, and **pterygoids**—plus these four: (1) **anterior belly of digastric**, (2) **mylohyoid**, (3) **tensor veli palatini**, and (4) **tensor tympani**, attaching to the malleus.

The cartilage bar of the second arch, Reichert's cartilage, is associated with the development of the upper part of the **hyoid bone**, including its lesser cornua (plural of *cornu*, Latin for "horn"), the **styloid process, stylohyoid ligament**, and the third ear ossicle, *i.e.*, the **stapes**. The resident motor nerve of this arch is N. VII. The facial nerve supplies the following branchiomeric muscles (the term *branchiomeric* refers to striated muscle originating in pharyngeal arch mesenchyme): all **muscles of facial expression**, including those that move the scalp, plus these four: (1) **posterior belly of digastric**, (2) **stylohyoid**, (3) **buccinator**, the muscle forming the substance of the cheek—more a muscle of mastication than of facial expression, and (4) **stapedius**, attaching to the stapes.

The lower part of the hyoid bone, including the greater horns, comes from the cartilage bar of the third pharyngeal arch. One muscle, and only one muscle, **stylopharyngeus**, develops in third arch mesenchyme; it is supplied by N. IX, the resident nerve of the arch.

The cartilages of the remaining pairs of pharyngeal arches turn into **laryngeal cartilages**—thyroid, arytenoid, and cricoid; and arch mesenchyme is the source of the striated muscle of **pharynx** and **larynx**. The superior and recurrent laryngeal branches of N. X are resident nerves of arches four and six, respectively. Sternocleidomastoid and trapezius are assumed to be branchiomeric muscles as well; both are innervated, of course, by the spinal root of N. XI.

An aortic arch artery runs through the substance of a same-numbered pharyngeal arch. As previously discussed, the aortic arch arteries belonging to the third pair of pharyngeal arches persist as internal carotid arteries; those of the fourth pair of pharyngeal arches become arch of the aorta on the left, and right subclavian artery on the right. The sixth pair of aortic arch arteries associated with the last pair of pharyngeal arches remains as pulmonary arteries and, on the left, as the ductus arteriosus.

The Triangles of the Neck

Figure 12.5 a–b illustrates the limits of anterior and posterior cervical triangles. The base of the **anterior triangle** is the lower margin of the mandible, and its sides are the anterior borders of sternocleidomastoid muscles. Each **posterior triangle** has the intermediate one-third of the clavicle for a base and trapezius and sterncleidomastoid muscles for its sides.

The large anterior triangle is divisible into smaller right and left anterior triangles by an imaginary vertical line connecting the midpoint of the jaw with the upper border of the manubrium. The right and left anterior triangles, in turn, are partitioned by the bellies of the digastric muscles above and the superior belly of the omohyoid muscle below; and each posterior triangle is subdivided by the inferior belly of omohyoid into two triangles of unequal size. Their names provide clues as to contents.

The Transcricoidal Plane

Now, imagine a horizontal plane passing through the neck at the level of the cricoid cartilage (Fig 12.6). The cricoid is a signet-ring shaped laryngeal cartilage connected to the thyroid cartilage above by the cricothyroid ligament. To locate the cricoid cartilage on the body's surface, find the laryngeal prominence, or "Adam's apple." Inferior to it is a shallow groove whose floor is the median cricothyroid ligament, and just below that is the arch of the cricoid cartilege.

The **transcricoidal plane** intersects the cervical spine behind at vertebral level C6. C6 is the last cervical vertebra to have a bifid and non-palpable spinous process. That of vertebra C7, the vertebra prominens, is undivided.

Here are a few more comments about vertebra C6: the anterior tubercles on its transverse processes (Chassaignac's tubercles) are the largest of any cervical vertebra. They carry the proprietary designation of *carotid tubercles*, for the common carotid arteries may be compressed against them, cutting off the blood supply to the brain and causing loss of consciousness.

The cricoid cartilage, not unlike the sternal angle, is a palpable and fairly stable landmark that can guide you to several points of interest situated on or about the transcricoidal plane. Keep in mind, however, that the cricoid, being part of the larynx, will change its position relative to the vertebral column; *e.g.*, during swallowing or vocalization. Nevertheless, you may associate vertebral level C6 with: (1) the beginning of the trachea and the end of the larynx, (2) the beginning of the esophagus and the end of the pharynx, and (3) the beginning of the vertebral artery's ascent through transverse foramina (it skips C7), and the end of the vertebral vein's bony passage. Located on or about the transcricoidal plane also are (4) the middle cervical ganglion of the sympathetic trunk, (5) the loop at the bottom of the ansa cervicalis buried in the carotid sheath, (6) the loop of the inferior thyroid artery, giving rise to (7) the ascending cervical artery, and (8) the middle thyroid vein entering the internal jugular vein. Add to the list, (9) the recurrent laryngeal nerves which enter the larynx by passing up and under the lower border of the inferior constrictor muscle. Further, do not forget that (10) the vocal folds are just above the transcricoidal plane, and that (11) the isthmus of the thyroid gland is only a short distance below it, straddling tracheal cartilage rings two, three, and four. Indeed, (12) the superior parathyroid glands embedded in thyroid gland tissue lie close to the transcricoidal plane.

The Cervical Fascia

Take a look now at the arrangement of the deep fascia of the neck in horizontal sections (Fig. 12.7 a–d). Its superficial or **investing layer** splits to enclose sternocleidomastoid and trapezius muscles (Fig. 12.7 a), both innervated by N. XI. The part of the investing fascia separating sternocleidomastoid from trapezius is the roof of the posterior cervical triangle. It is coarsely fibrous, especially so posterolaterally, and stubbornly resists the dissector's best efforts.

The **prevertebral layer** (Fig. 12.7 b) is strong and well-developed over the prevertebral muscles. Its lateral extension covers paravertebral muscles and forms the floor of the posterior cervical triangle. Beneath it, and relatively secure for as long as the fascia is intact, are the phrenic nerve and muscular branches of the brachial plexus. Notice that the sympathetic trunk is outside the prevertebral layer.

The **visceral fascia** (Fig. 12.7 c) is in contact with the trachea in front, the thyroid gland laterally, and the esophagus behind. Its pretracheal layer, being the upward extension of pericardium, is strong. The layer of visceral fascia separating the esophagus from the prevertebral region, however, the **buccopharyngeal fascia**, is thin and non-descript; *i.e.*, it is hardly deserving of special name or consideration. But, then again, the layer is the anterior boundary of the **retropharyngeal space** through which infections may spread into the posterior mediastinum, or from one side of the neck to the other.

The **carotid sheath** of cervical fascia, (Fig. 12.7 d) ties the common carotid artery, internal jugular vein, and vagus nerve into a neurovascular bundle. It is densely fibrous, thereby offering a fair amount of protection for their passage through the neck.

Finally, the deep fascia of the neck clothing the infrahyoid muscles is singled out by some authors as the **depressor hyoid fascia**. A few even go so far as to specify superficial and deep layers. Perhaps the more critical issue here is that this division of the cervical fascia creates tissue spaces that are continuous below with superior and anterior mediastina.

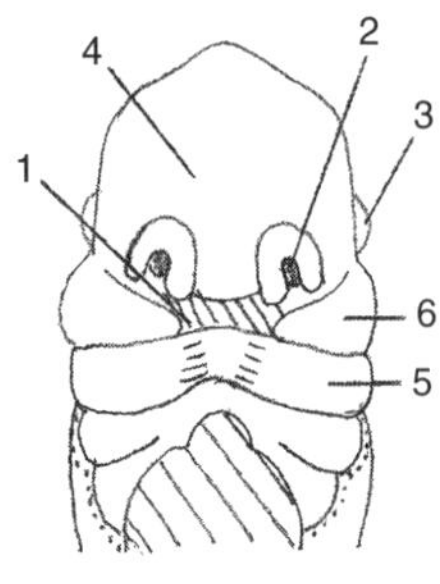

12.1 Facial development

1. Oropharyngeal membrane
2. Nasal pit
3. Eye
4. Frontonasal process
5. Mandibular process of 1st arch
6. Maxillary process of 1st arch

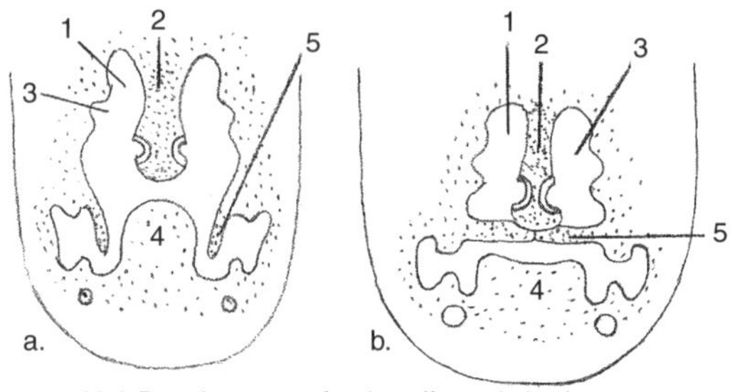

12.2 Development of palate (frontal view)

a. Early stage b. Late stage

1. Nasal cavity
2. Nasal septum
3. Paranasal sinus
4. Tongue
5. Palatine process of maxilla

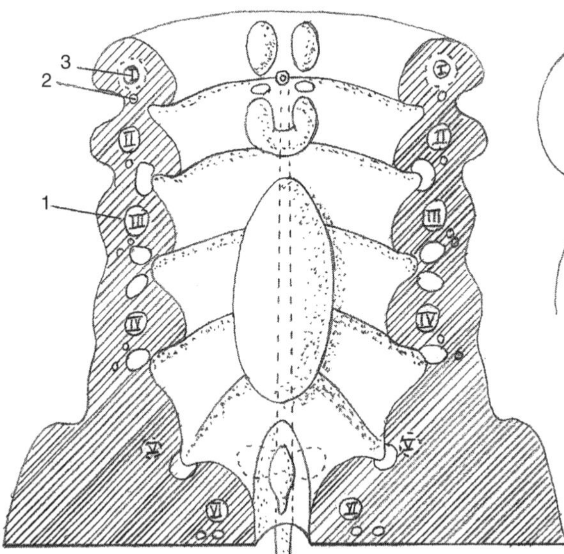

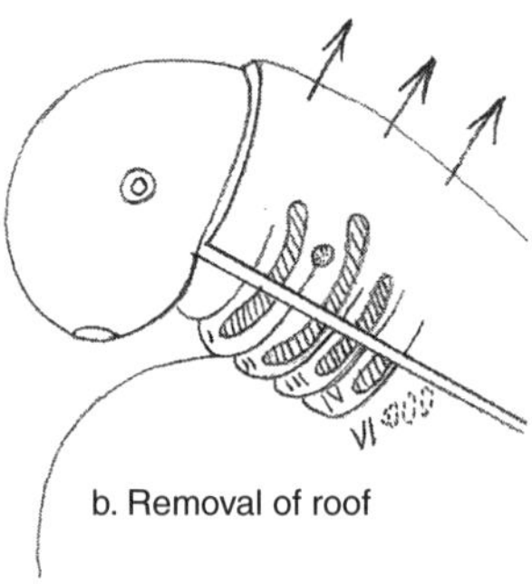

b. Removal of roof

12.3a Primitive pharynx

a. Floor and sides with roof removed (dorsal view)

1. 3rd arch with int. carotid a. and glossopharyngeal n.
2. Mandibular division of trigeminal n.
3. Meckel's cartilage

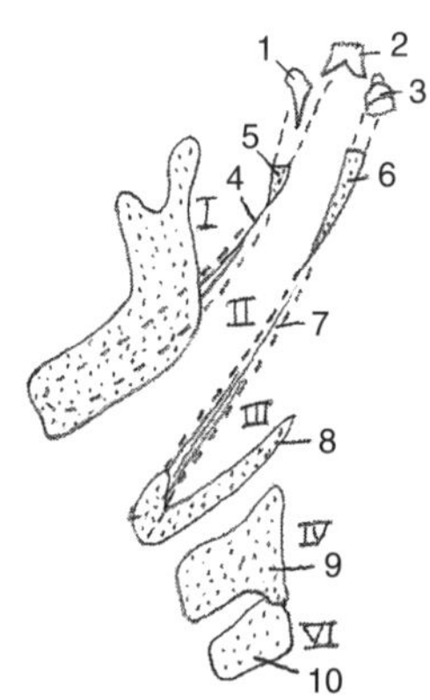

12.4 Fate or arch cartilages

1. Malleus
2. Incus
3. Stapes
4. Sphenomandibular lig.
5. Spine of sphenoid bone
6. Styloid process
7. Stylohyoid ligament
8. Hyoid bone
9. Thyroid cartilage
10. Cricoid cartilage

Figure 12.1: Development of the face
Figure 12.2: Development of the palate
Figure 12.3a-b: The structure and derivatives of pharyngeal arches
Figure 12.4: Fate of arch cartilages

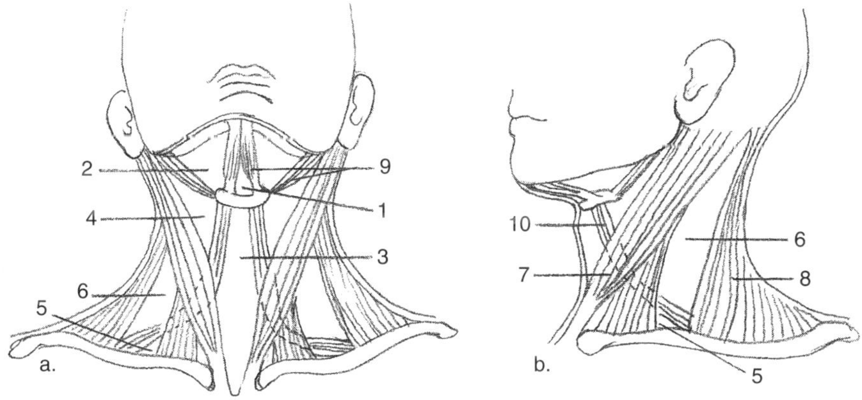

12.5a-b Triangles of neck

1. Submental
2. Submandibular
3. Muscular
4. Carotid
5. Supraclavicular
6. Occipital
7. Sternocleidomastoid m.
8. Trapezius m.
9. Digastric m.
10. Omohyoid m.

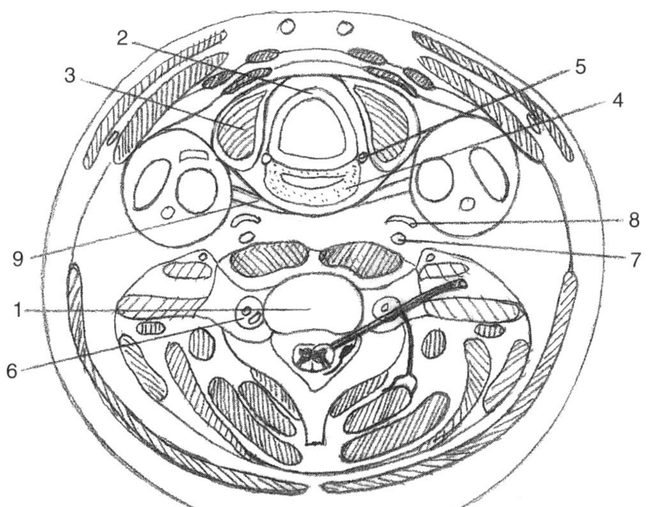

12.6 Transcricoidal plane

1. Body of vertebra C6
2. Beginning of trachea
3. Thyroid gland
4. Beginning of esophagus
5. Inferior laryngeal nerve
6. Vertebral vessels
7. Middle cervical ganglion
8. Loop of inferior throid artery
9. Middle thyroid vein

Figure 12.5a-b: Triangles of neck
Figure 12.6: Transcricoidal plane

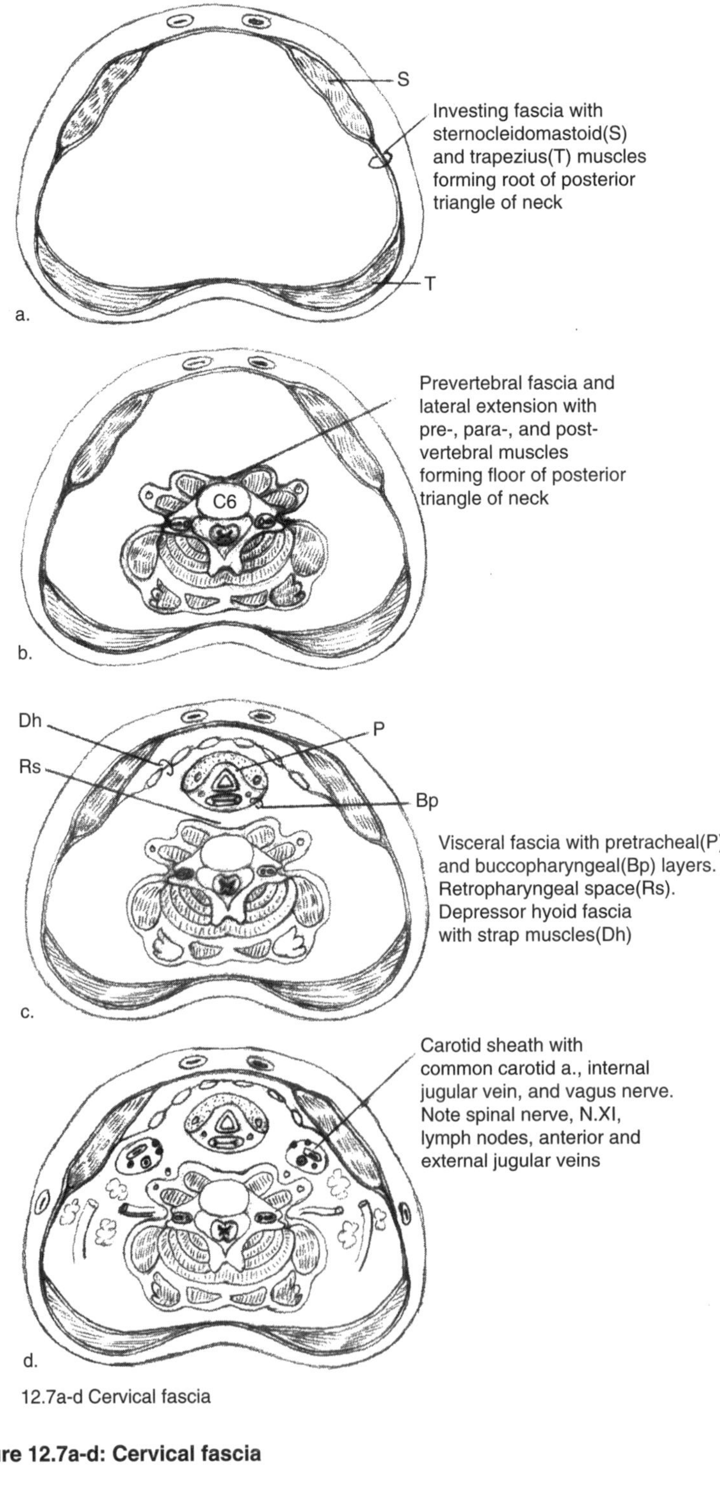

12.7a-d Cervical fascia

Figure 12.7a-d: Cervical fascia

Chapter 13

INTRODUCTION TO ANGIOLOGY

Angiology is the study of the bloodvascular system—its evolution (angiogenesis), its patterns of distribution, and its central function of protecting the body from ischemic disease. A major characteristic of the circulatory system is its versatility and redundancy: most arteries or veins carrying blood to and from regions of the body have overlapping distributions, and virtually all communicate or establish connections (anastomoses) as need dictates.

Angiogenesis

Blood vessels begin to form during the third week of embryonic life, in the late presomite stage. As discussed in chapter 10, their development is preceded by the appearance of angioblastic cords in the mesoderm of the cardiogenic area. The cords canalize to become endothelial tubes, the most prominent of which are the paired, primitive dorsal aortae. Angioblastic cords are found additionally in extra-embryonic locations, *e.g.*, in the yolk sac where they give rise to the **vitelline arteries** and **veins**, also in the body stalk and the chorion. It is likely that the vaso-formative cells in the body stalk were brought there by mesoderm surrounding the **allantoic diverticulum**, a caudal extension of the yolk sac cavity that invades the body stalk towards the middle of the third week (see Fig 10.5, p. 66). The allantois soon degenerates, but the allantoic vessels persist. They are later renamed **chorionic**, then **umbilical arteries** and **veins** and carry fetal blood to and from the placenta.

Continuing endothelial growth generates an extensive, dense web of vessels inside the embryo, but only a small part of this primitive vascular bed is utilized in the establishment of named arteries and veins. Which parts are chosen, and which are not, depends on genetic predisposition, mechanical forces aiding or resisting blood flow, and cardiovascular hemodynamics.

Angiogenesis is an important process not only before, but also after birth: wound-healing and tissue repair at all ages are dependent on endothelial regeneration. It is, however, not always a restorative process: tumor growth and survival, for instance, are correlated with vascular proliferation and increased blood flow (hyperemia) to the tumor site. Small wonder, then, that the search for angiogenesis inhibitors is an active area of cancer research.

The Vascular Backup System

The primitive vascular bed remains intact throughout the lifetime of an individual, but only parts of it are fully engaged in carrying blood through the adult cardiovascular system. The rest persists as a pool of substitute or replacement vessels that can be utilized in times of circulatory distress; *i.e.*, when blood flow through primary channels is obstructed or interrupted.

Cessation of blood flow to a region may occur spontaneously, as it does in the presence of a clot; or, it may be deliberately planned as part of some surgical procedure, when vessels must be tied and divided. Vascular compression syndromes reduce blood flow by narrowing a vessel's lumen. Whatever the nature of the crisis, the system will respond in one or more ways—none of which, however, entails *de novo* creation of blood vessels! One compensatory mechanism is renewed growth and elongation of preexisting capillaries, a second is the recruitment of reserve vessels from the original pool. Of course, both mechanisms may operate simultaneously.

Now, the process of capillary growth and elongation takes time, perhaps weeks; and, while it may work when there is gradual reduction in blood flow, it is not a practical remedy to counter the effects of sudden vascular obstructions. A much more effective, efficient, and immediate solution is to open up the lines of inactive vessels that were laid down early in development. "Resting" capillaries, by dilating and/or by fusing with each other, may carry a vascular load past a clot or ligature, thus ensuring survival of oxygen-deprived tissues. Such rerouting of blood is the basis of **collateral circulation**, which, loosely defined, refers to the vascular circuitry that allows blood to perfuse an organ or area when its principal source is blocked or no longer equal to the task.

Collateral Circulation

All arteries, with very few exceptions, communicate with other arteries by means of precapillary anastomoses. The connections are made usually by distant and diffuse branches of two or more small arteries having overlapping distributions and, therefore, supplying the same, or closely related, parts of the body. Postcapillary connections are established by distant and diffuse tributaries of small veins.

In general, vessels lacking the capacity to establish adequate collateral circulation in short order are **functional end arteries**. Abrupt cessation of blood flow in such vessels almost invariably has catastrophic results—as does, for example, thrombosis of the lenticulostriate arteries supplying the brain, or sudden occlusion of coronary arteries. When, however, the reduction in blood flow is gradual, and the communications between collateral vessels are abundant, the risk of irreversible damage downstream from an obstruction is diminished. Indeed, cases have been reported of individuals having led reasonably normal lives with severely clogged coronary arteries.

And, speaking of **coronary arteries**, let me remind you that they are the first vessels off the aorta, and that blood is forced into them during ventricular diastole while the aortic semilunar valve is closed and blood in the aorta rebounds. The right coronary artery is the dominant vessel in about 90 percent of cases, meaning that is supplies the right side of the heart *plus* a portion of the left ventricle on the diaphragmatic aspect of the heart. The artery does so via its posterior interventricular branch. In about ten percent of cases, the posterior interventricular branch comes off the circumflex artery, in which case the left coronary vessel is said to be dominant.

Arteries that completely lack vascular connections are known as **true end arteries**. Fortunately, such anatomically isolated vessels are exceedingly rare; for, happenstances that abolish blood flow in these vessels cause irreversible damage in their fields of distribution. The central artery of the retina is the prime example of a true end artery—its blockage causes permanent blindness.

Collateral Circulation of Head and Neck

As noted in an earlier chapter, regions such as the chest wall that receive blood from a variety of sources are vascular hubs: arteries communicate freely with one another, thereby allowing blood to be switched between and among vessels. The largest and by far the best developed arterial anastomosis, however, is at the base of the brain—the cerebral arterial circle (of Willis), a center-piece of collateral circulation designed to maintain blood flow to the brain at all times.

The blood supply to head and neck is built around two large arterial systems, carotid and subclavian, which anastomose by a variety of routes. The carotid system begins with the **common carotid arteries** which have no branches other than their terminals. Each divides, roughly opposite vertebra C4, into external and internal carotid arteries. External carotid arteries supply parts of the neck and most of the head external to, but including, the cranial dura. Internal carotid vessels have no branches in the neck.

Each **external carotid artery** terminates by dividing behind the ramus of the mandible into maxillary and superficial temporal arteries. Each has altogether eight named branches (Fig. 13.1): four vessels are directed medially (**superior thyroid, lingual, facial,** and **maxillary**), two head cranially (**ascending pharyngeal** and **superficial temporal**), and two pass posteriorly (**occipital** and **posterior auricular**). Branches of right and left external carotid arteries anastomose close to the midline; collateral circulation between contralateral sides is thereby assured.

The **internal carotid arteries** enter the cranial cavity through the carotid canals and feed into the Circle of Willis (Fig. 13.2). Their terminals are **anterior** and **middle cerebral arteries**. Before joining the arterial circle, each internal carotid artery gives off an **ophthalmic artery**, which passes through the optic canal into the orbit. Its many branches include the **central artery of the retina**. The ophthalmic artery is a key vessel: it not only supplies all structures inside the orbit, but also anastomoses with branches of the external carotid artery. The **supra-orbital branch** of the ophthalmic artery, for example, makes connections in the superficial fascia of the scalp with the superficial temporal branch of the external carotid, and the **dorsal nasal branch** meets the facial artery outside the orbit. Thus, blood is free to circulate between all parts of the carotid system.

The **subclavian system** can be extended into the axilla, for the subclavian artery continues as **axillary artery** into the arm. The branches of this system involved in head and neck anastomoses are (1) the **vertebral arteries**, (2) the **thyrocervical trunks** and, to a lesser extent, (3) the **costocervical trunks** (Fig. 13.3). The vertebral vessels, the first branches off the subclavian arteries, unite to form the single **basilar artery**. Its terminal branches, the **posterior cerebral arteries**, link up with the internal carotid system to complete the Circle of Willis. The paired thyrocervical trunks communicate with the carotid system on a number of levels: their **inferior thyroid branches** anastomose with the superior thyroid branches of the external carotids in the substance of the thyroid gland—creating another arterial circle of sorts; and their **transverse** and **ascending cervical vessels** make contact with ascending pharyngeal, occipital, and posterior auricular branches of the external carotid arteries. The **deep cervical branches** of the costocervical trunks distribute in a similar manner. The net effect is an abundance of collateral pathways between subclavian and carotid systems.

The base of the brain is not the only place inside the cranial cavity where carotid and subclavian systems intersect. Though mainly distributing to bone, **anterior meningeal branches** of the ophthalmic arteries, **middle meningeal branches** of the maxillary arteries, and **posterior meningeal branches** of the vertebral arteries anastomose on the cranial (periosteal) dura. Also, let us not forget the scalp, where internal and external carotid arteries are prominently and bilaterally tied together through supra-orbital, supratrochlear, superficial temporal, posterior auricular, and occipital branches. The subclavian vessels participate in this massive anastomosis by virtue of their cervical branches, which communicate with both posterior auricular and occipital branches of the external carotid arteries. No wonder, then, that open scalp wounds bleed profusely and are difficult to close.

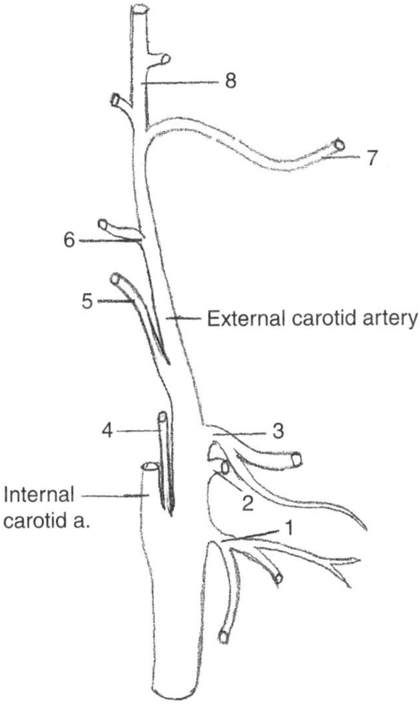

1. Superior thyroid artery
2. Lingual artery
3. Facial artery
4. Ascending pharyngeal artery
5. Occipital artery
6. Posterior auricular artery
7. Maxillary artery
8. Superficial temporal artery

13.1 External carotid system

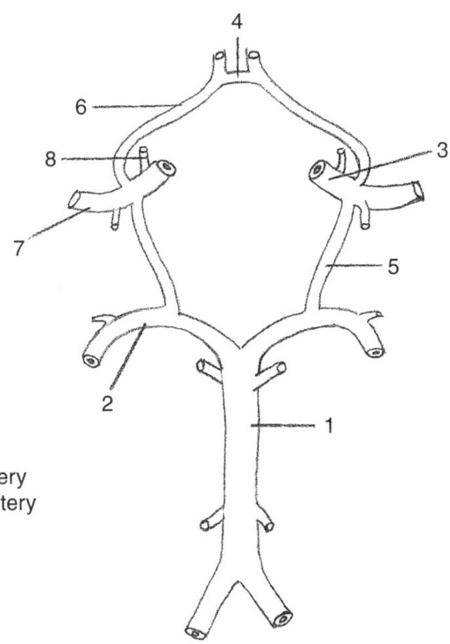

1. Basilar artery
2. Posterior cerebral artery
3. Internal carotid artery
4. Anterior communicating artery
5. Posterior communicating artery
6. Anterior cerebral artery
7. Middle cerebral artery
8. Ophthalmic artery

13.2 Cerebral arterial circle (of Willis)

Figure 13.1: External carotid system
Figure 13.2: Cerebral arterial circle (of Willis)

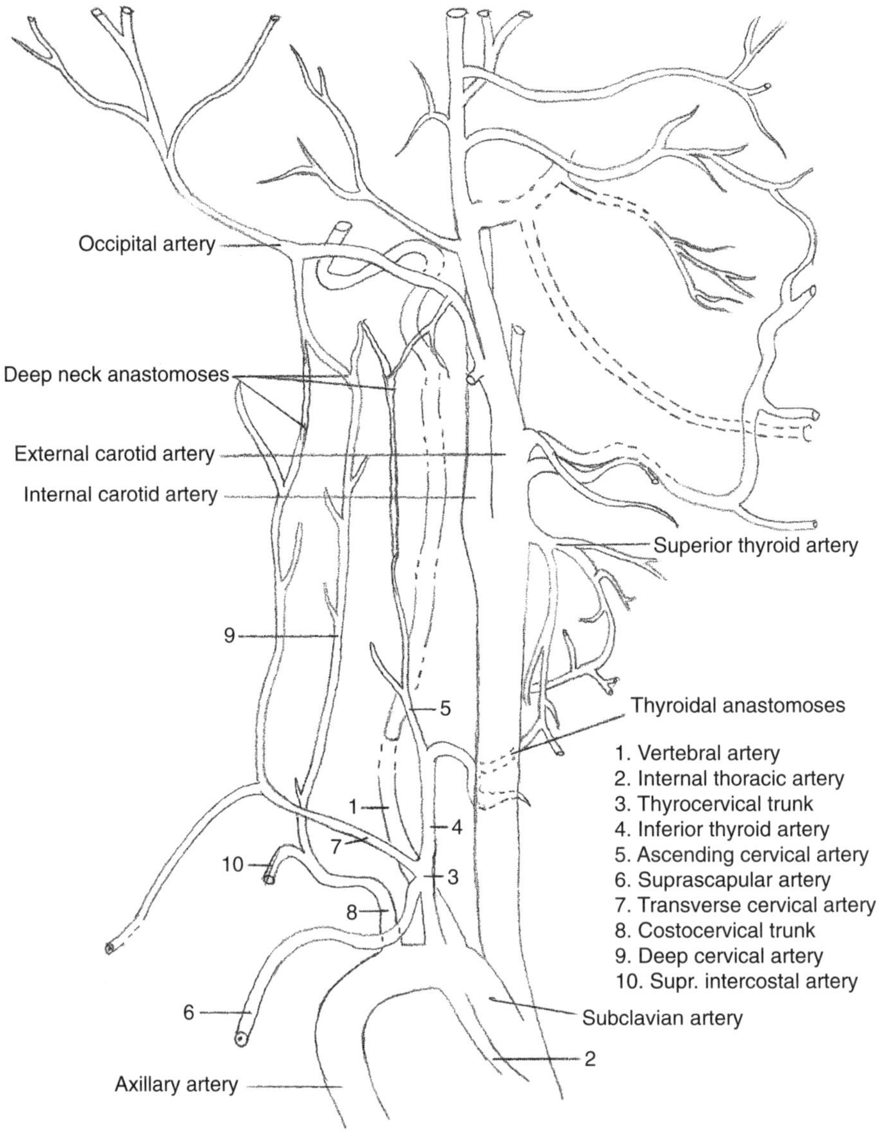

13.3 Arterial neck anastomoses

Figure 13.3: Arterial neck anastomoses

Chapter 14
AUTONOMIC INNERVATION OF HEAD AND NECK

Innervation to the tissues of the head and neck is provided almost exclusively by 12 pairs of cranial nerves. As you are well aware by now, cranial nerves are an integral part of the peripheral nervous system (PNS), which transmits signals to and from the central nervous system (CNS). Some of the cranial nerves carry parasympathetic fibers of the autonomic nervous system (ANS); others contain fibers that supply branchiomeric muscle.

Like other peripheral nerves, cranial nerves have sensory and/or motor fibers, classifiable as either somatic or visceral.

Classification of Sensory Fibers

General somatic sensory fibers carry impulses from the skin, including its appendages (*e.g.*, hair follicles), and from specialized surface receptors for modalities such as temperature or touch—*i.e.*, they transmit *exteroceptive* stimuli. Somatic sensory fibers also carry messages from receptors in tendons and muscle; they transmit muscle sense, a modality known as *proprioception*—a sense of the position of the body in space.

General visceral sensory fibers bring messages from internal organs to the CNS, thus transmit *visceroceptive* or *enteroceptive* signals.

Certain cranial nerves contain fibers that carry input from organs of special sense; *e.g.*, the retina, olfactory epithelium, or taste buds. Such fibers are **special** sensory fibers, and they are found exclusively in NN. I, II, VII, VIII, IX, and X. Many textbooks classify sensory fibers of cranial nerves subserving feeding behavior, including smell and taste (NN. I, VII, IX, X), as special visceral sensory fibers and all others as special somatic sensory fibers.

Classification of Motor Fibers

Somatic motor fiber distribute to skeletal, voluntary muscle. **General visceral** motor fibers innervate the heart and smooth muscle, including the arrector pili bundles in the dermis. They supply the smooth muscle in the walls of blood vessels, thus exercising vasomotor control, and are secretomotor to salivary and lacrimal glands. In short, they innervate parts of the body that function involuntarily. We have now arrived at a definition of the **autonomic nervous system** (ANS), to wit: it is the sum total of general visceral motor fibers in the body that supply excitable tissue over whose actions we normally have scant control. More about the ANS in a moment.

Certain cranial nerves contain fibers that distribute to striated muscle of head and neck developed not from somites but from head (pharyngeal arch) mesoderm. They are classified as **special** motor fibers and innervate the muscles of mastication (N. V), facial expression (N. VII), pharynx and larynx (NN. IX–XI). Again, many textbooks, citing the feeding-related functions of branchiomeric muscles, refer to these fibers as special visceral motor fibers.

The Autonomic Nervous System

The ANS, by convention, is defined as strictly a motor system, for visceral sensory fibers do not differ from somatic sensory fibers of the PNS, except in that they transmit visceroceptive rather than exteroceptive or proprioceptive messages. General visceral motor fibers, however, differ conspicuously from somatic motor fibers both structurally and functionally.

At least six cranial nerves distribute general visceral motor fibers to parts of the head and neck and, thus, are affiliated with the ANS. To place them in proper context, a brief review of the ANS may be in order.

Signals travelling over the ANS reach their destination via a two-neuron chain. The cell bodies of first-order neurons are located in the CNS. Their axons terminate by synapsing with second-order neurons whose cell bodies form ganglia outside the CNS (Fig. 14.1). Thus, each general visceral motor nerve has preganglionic and postganglionic parts.

The function of the ANS is to regulate the *milieu interieur*, the term used by the distinguished French physiologist, Claude Bernard (1813–1878), to describe the body's internal environment. The preeminent American physiologist, (1871–1945), referred to *homeostasis* as the natural tendency of the body to maintain stable levels of function. The ANS, operating through its sympathetic and parasympathetic divisions, maintains the constancy of the internal environment and reacts to stresses, internal or external, to which it may be subject.

The **parasympathetic division** is predominantly active when the body is at rest. It slows the heart rate, aids the digestive process, and stimulates peristalsis. It relaxes all involuntary sphincters, *except* m. sphincter pupillae which contracts in response to parasympathetic excitation.

The **sympathetic division** of the ANS energizes the body for "fight or flight," to use Cannon's descriptive terms. Blood pressure rises, heart rate increases, and bronchioles dilate. It causes sudomotor (sweat-gland) and pilomotor (arrector pili) activity, and stimulates adrenomedullary secretion.

The two divisions of the ANS differ from each other not only functionally but also structurally. They differ with regard to (1) their segmental outflow from the CNS, (2) location of their ganglia, and (3) the relative lengths of their pre- and postganglionic fibers.

Cell bodies of sympathetic first-order neurons are present only in thoracic spinal cord segments T1–12 and lumbar segments L1–2; those of the parasympathetic system reside in the brain stem and in sacral spinal cord segments S2–4. Hence, thoracolumbar outflow equates with sympathetic division, and craniosacral outflow with parasympathetic division.

The ganglia of the sympathetic trunk contain the cell bodies of second-order sympathetic neurons. The paired sympathetic trunks start at the base of the skull, run the length of the vertebral column, and join over the coccyx to form the ganglion impar—the "unpaired" ganglion (Fig. 14.2). Some 23 pairs of sympathetic ganglia lie in between: superior, middle, and inferior cervical ganglia in the neck, 11–12 ganglia in the thoracic region, and four ganglia each in the lumbar and sacral regions. Because of their location close to the vertebral column, they may be classified as sympathetic paravertebral ganglia. Prevertebral sympathetic ganglia surround the celiac, superior, and inferior mesenteric arteries and are named accordingly.

Discreet parasympathetic ganglia containing second-order neurons are found only in the head region. I shall discuss the parasympathetic head ganglia and their connections shortly. The rest of the second-order neurons relaying parasympathetic signals are located near or within the organs for which they are targeted.

As sympathetic ganglia lie closer to the spinal cord than do parasympathetic ganglia, preganglionic sympathetic fibers are shorter than preganglionic parasympathetic fibers; the reverse is true for postganglionic fibers. Preganglionic sympathetic fibers leave the spinal cord via anterior roots to become part of a segmental spinal nerve. They exit the nerve almost at once, however, and as **white rami communicantes** enter the sympathetic ganglion of that segment (Fig. 14.1). Here, they may synapse with one or more cell bodies of second-order neurons; or, they may pass uninterruptedly through the ganglion to synapse in a ganglion of a different segment. Remember, although sympathetic fiber *outflow* is confined to thoracolumbar segments, the sympathetic trunk runs the length of the vertebral column. This arrangements allows preganglionic sympathetic fibers to ascend as high as the superior cervical ganglion and to descend as low as the ganglion impar. For the same reason, postganglionic sympathetic fibers may join, in the form of **grey rami communicantes**, spinal nerves of any segment. A third group of preganglionic sympathetic fibers passes through the sympathetic trunk without pause, bound for prevertebral ganglia. Running with general visceral sensory fibers, they form the greater

(T5–9), lesser (T9–11), and least (T12) **splanchnic nerves**. Postganglionic fibers issuing from the ganglia distribute to abdominal and pelvic organs.

The Parasympathetic Head Ganglia

There are four pairs of autonomic ganglia in the head, all part of the parasympathetic division of the ANS. None is readily accessible to the untrained dissector, for the ganglia are small and placed deeply in the recesses of the skull: the **ciliary ganglion** inside the orbit, the **pterygopalatine ganglion** in the fossa of the same name, the **otic ganglion** in the infratemporal fossa, and the **submandibular/sublingual ganglia** below the tongue.

The four head ganglia contain cell bodies of second-order neurons whose axons are postganglionic parasympathetic fibers. The cell bodies of first-order neurons, whose axons are the preganglionic parasympathetic fibers in the two-neuron chain, reside in the brain stem (Fig. 14.3). They form (1) the **Edinger-Westphal nucleus**, (2) the **superior** and **inferior salivatory nuclei**, and (3) the **posterior nucleus of the vagus nerve** (the term *nucleus* refers to a collection of nerve cell bodies inside the CNS).

Preganglionic parasympathetic fibers reach second-order neurons by attaching themselves to specific cranial nerves as those nerves are emerging from the brain stem, and are carried by them and their branches to the appropriate head ganglion. Thus, the preganglionic fibers whose cell bodies make up the Edinger-Westphal nucleus travel with N. III, which carries them into the orbit and to the ciliary ganglion. Those of the superior salivatory nucleus pass with N. VII to the pterygopalatine fossa and ganglion, using the greater petrosal nerve as means of transport; others accompany the chorda tympani and are brought to the submandibular region and the submandibular/sublingual ganglia. The preganglionic fibers arising in the inferior salivatory nucleus are conveyed to the infratemporal fossa and the otic ganglion via the tympanic branch of N. IX (Jacobsen's nerve), tympanic plexus, and the lesser petrosal nerve. And preganglionic parasympathetic fibers whose cell bodies are in the posterior nucleus of N. X travel with cervical, thoracic, and abdominal branches of the vagus nerve.

The parasympathetic head ganglia, in short, are relay stations for secretomotor fibers to lacrimal, salivary, and mucus-producing glands, and for near-vision accomodation.

Can you figure out how postganglionic parasympathetic fibers distribute to their target tissues, *i.e.*, the smooth muscle of the eyeball, the lacrimal, mucus-secreting, and salivary glands? Quite simply, they, too, "hitch-a-ride," metaphorically speaking, using branches of the one cranial nerve that goes to virtually every corner of the head—namely, the trigeminal nerve. Passing by, or through, parasympathetic head ganglia, divisions or branches of N. V pick up postganglionic parasympathetic fibers and deliver them to their destination: the ophthalmic division through its short ciliary nerves to the smooth muscle of the eyeball and via its lacrimal branch to the lacrimal gland; the maxillary division through its many branches, *e.g.*, zygomatic, nasopalatine, pharyngeal, to mucus-producing tissue of nose, mouth, and pharynx; and the mandibular division by way of lingual and auriculotemporal branches to the salivary glands. Be aware, however, and this is important, that the branches of N. V carrying postganglionic parasympathetic fibers *are functionally independent of the ANS*! All are sensory nerves transmitting messages to the CNS; their cell bodies are located in the trigeminal (semilunar) ganglion.

Sympathetic Innervation of the Head and Neck

Unlike parasympathetic nerves, the sympathetic fibers perform no secretomotor functions in the head or neck. Their main activity is vasomotor control and innervation of the dilator pupillae muscle.

The cell bodies of first-order sympathetic neurons supplying the head and neck are stationed in the upper thoracic segments of the spinal cord. Their axons, preganglionic sympathetic fibers, carry impulses to superior, middle, or inferior cervical ganglia where they establish synaptic connections with cell bodies of second-order neurons. Postganglionic sympathetic fibers, emerging from the ganglia, attach themselves in plexiform fashion to vessels, principally carotid arteries and their branches, supplying regions targeted for sympathetic innervation. Thus, postganglionic sympathetic fibers forming the **internal carotid plexus** are brought into the orbit by the ophthalmic branch of that artery. Consider also Vidian's nerve, the **nerve of the pterygoid canal**, which passes into the pterygopalatine fossa. Its components are (1) preganglionic parasympathetic fibers of the greater petrosal nerve, and (2) the **deep petrosal nerve**—postganglionic sympathetic fibers separating from the internal carotid plexus as the artery begins its intracranial course. Likewise, postganglionic sympathetic fibers of the external carotid plexus are "piggy-backed" by that artery and its branches into the infratemporal fossa and the submandibular region.

Be certain you understand that postganglionic sympathetic fibers are *functionally independent of parasympathetic head ganglia*! They run straight past the head ganglion; and, even when they seem to be connected to one, postganglionic sympathetic fiber never synapse in it. Like postganglionic parasympathetic fibers, however, they, too, utilize branches of the trigeminal nerve to reach their destination. Indeed, most of the time postganglionic sympathetic and parasympathetic fibers keep each other company.

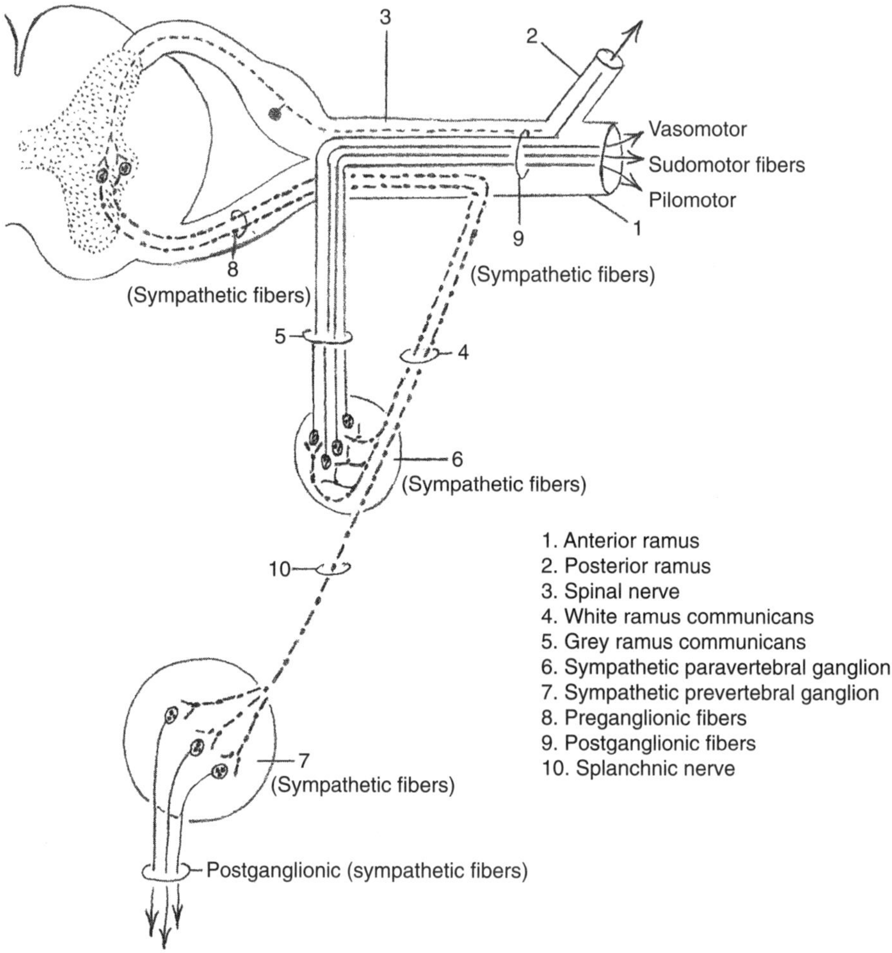

14.1 Distribution of a sympathetic nerve

Figure 14.1: Distribution of a sympathetic nerve

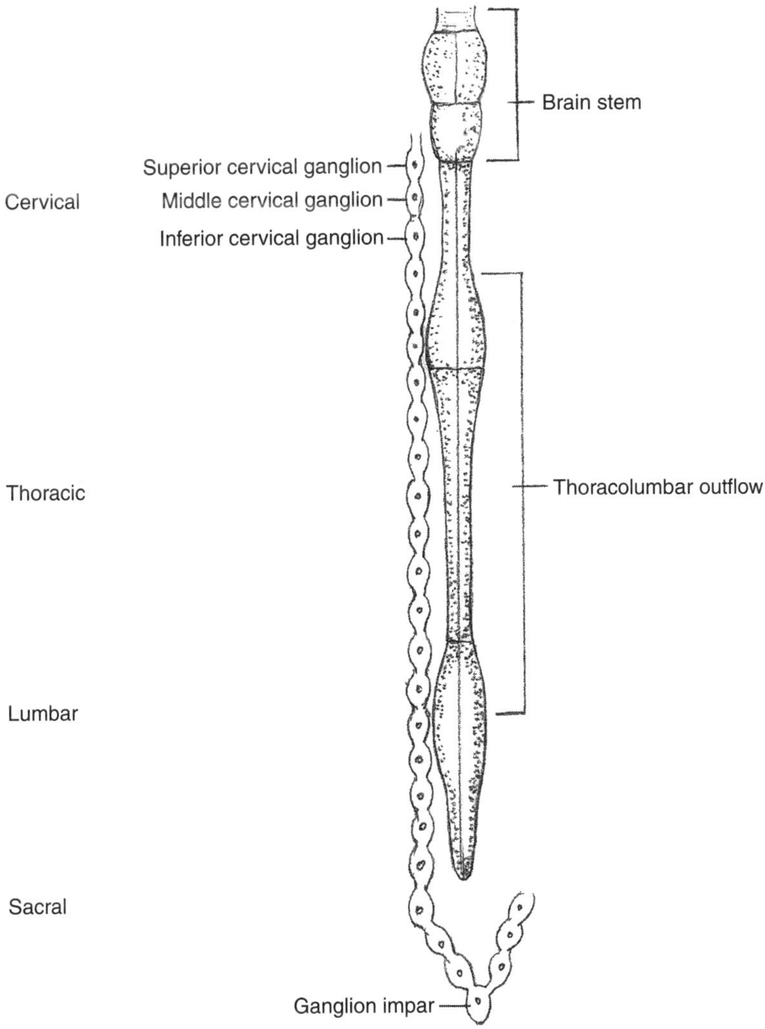

14.2 Spinal cord and sympathetic trunk

Figure 14.2: Spinal cord and sympathetic trunk

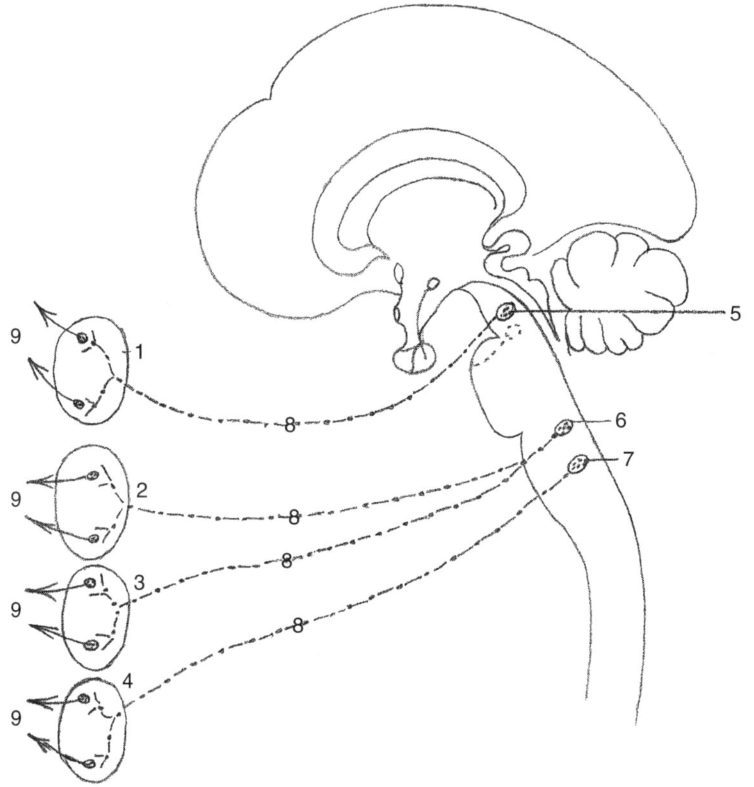

1. Ciliary
2. Pterygopalatine
3. Submandibular/sublingual
4. Otic
5. Edinger-Westphal nucleus
6. Superior salivatory nucleus
7. Inferior salivatory nucleus
8. Preganglionic parasympathetic fiber
9. Postganglionic parasympathetic fiber

14.3 Parasympathetic head ganglia

Figure 14.3: Parasympathetic head ganglia

Chapter 15
THE PHARYNX

The pharynx (pronounced *phar-ynx*) is a fibromuscular passage utilized by both the respiratory and the digestive systems. Being embedded deeply in the neck, the pharynx is a relatively inaccessible organ; hence, its shape, structure, and overall place in the scheme of things are a mystery to many.

The pharynx attaches above to the base of the skull and is continuous below, opposite vertebra C6, with the esophagus. Unlike the esophagus, however, the pharynx lacks an anterior wall where it is in direct and open communication with the back of the nasal and oral cavities. It always reminds me of an oxygen or gas mask turned backwards.

Below and behind the tongue, the pharynx is in back of the larynx (pronounced *lar-ynx*); thus, the pharynx is divisible into three parts: **nasopharynx, oropharynx,** and **laryngopharynx** (Fig. 15.1).

Nasopharynx

The nasopharynx is the part between the base of the skull and the palate; thus, the palate separates nasopharynx from oropharynx. The two parts become completely shut off from each other during deglutition, also during suckling, when **levator** and **tensor veli palatini muscles** elevate the soft palate until it makes contact with the posterior wall of the pharynx. The latter is simultaneously drawn forward by contracting fibers of the superior constrictor muscle; this prominent bulge is Passavant's **palatopharyngeal ridge**.

The pharyngeal opening of the **pharyngotympanic** (Eustachian or auditory) **tube** projects into the lateral wall of the nasopharynx just above the soft palate. Prominent anterior and posterior folds of mucous membrane frame the tube's entrance into the pharynx. The anterior fold of the tubal elevation or **torus tubarius** (L., *torus* = a swelling) is the salpingopalatine fold (Gr., *salpynx* = a tube); it blends with the soft palate below. The somewhat larger, salpingopharyngeal fold descends almost vertically from the posterior margin of the torus tubarius. It contains the salpingopharyngeus muscle, which mingles below with constrictor and palatopharyngeus muscles of the pharyngeal wall. A third mucosal fold, in the floor of the torus, overlies the levator veli palatini muscle (levator palati, for short), which accompanies the auditory tube into the nasopharynx.

Oropharynx

This part of the pharynx is located between the palate and the hyoid bone, thus lies opposite and anterior to vertebrae C1–3. In front, where soft palate and pharyngeal wall merge, the oropharynx communicates with the oral cavity through the **oropharyngeal isthmus**, or **fauces** (L., *fauces* = a passage).

The fauces (Fig. 15.2) is bounded laterally by the palatine arches, or pillars of the fauces—paired folds of mucous membrane containing striated muscle. Between each pair is a tonsillar sinus, home to the **palatine tonsil**. The posterior pillar of the arch, the palatopharyngeal fold and muscle, stretches from the soft palate above to the posterior border of the thyroid cartilage below. The anterior pillar of the arch, palatoglossal fold and muscle, extends from the soft palate to the posterolateral aspect of the tongue. The palatoglossus muscle, you will recall, is the only extrinsic muscle of the tongue *not* supplied by the hypoglossal nerve (N. XII); it is innervated by branches of the vago-accessory complex (N. X–XI). Also remember this: all but one of the muscles acting on the soft palate are innervated by branches of the pharyngeal plexus or vago-accessory complex. The exception is the tensor veli palatini muscle (tensor palati, for short), which develops from mesoderm of the first pharyngeal arch, hence is supplied by the trigeminal nerve (N. V).

The oropharyngeal isthmus overlies the sulcus terminalis of the tongue. Both are watersheds dividing territories having distinctly different sensory innervation: the mucous membrane in front of isthmus and sulcus is supplied by N. V; that behind by N. IX and, more distally, by N. X–XI.

Laryngopharynx

The laryngopharynx begins opposite the **laryngeal aperture** or **inlet** and ends at vertebral level C6, where it is continuous with the esophagus. This part of the pharynx, therefore, is situated behind the root of the tongue and related to the larynx posterolaterally (Fig. 15.3). Along the sides of the laryngeal inlet the pharynx widens to form the paired **piriform fossae** (L., *pirum* = a pear). Liquid and solid foods cascade over them during the second stage of deglutition; sometimes, small chicken bones or other bony fragments lodge in a piriform fossa and may have to be removed by an emergency-room physician.

The Pharyngobasilar Fascia

The wall of the pharynx, like the walls of virtually all hollow organs in the body, is composed of four concentric coats. The innermost coat is the mucosal layer, next to which is the submucosa. Muscle lies external to the submucosa, and the serosa forms the exterior coat.

The submucosal layer of the pharyngeal wall is well developed and may be regarded as the skeleton of the pharynx. It is especially strong in the upper nasopharynx where, as **pharyngobasilar fascia**, it connects the pharynx to the cranium by blending with the periosteum of bones at the base of the skull. The fascia sweeps from its attachment on the medial pterygoid plate of one side across the pharyngeal tubercle on the basilar part of the occipital bone to the medial pterygoid plate of the other side (Fig. 15.4). In its semicircular course, the pharyngobasilar fascia crosses the cartilaginous part of the auditory tubes and the petrous portion of the temporal bones just in front of the carotid canal. The surface projection of the pharyngobasilar fascia at its attachment to the base of the skull is in the same horizontal plane as the orbital floors, the external auditory meati, and the external occipital protuberance.

The Constrictor Muscles

The three paired **constrictor muscles**, superior, middle, and inferior, form overlapping sheets in the pharyngeal wall external to the pharyngobasilar fascia. Their fibers are predominantly horizontally disposed, and their squeezing action moves luminal contents into the esophagus below.

Right and left constrictors muscles pass from bony, cartilaginous, or ligamentous attachments laterally towards the back of the pharynx (Fig. 15.5). Posteriorly, in the midline, they are "stitched" together in a vertical seam, the **pharyngeal raphe**, that runs the length of the organ starting with the pharyngeal tubercle above. The pharyngobasilar fascia lies exposed, however, in the pharyngeal wall laterally where superior and middle constrictors, and middle and inferior constrictor muscles, do not overlap. Pharyngobasilar fascia remains uncovered also where the upper border of the superior constrictor fails to meet the skull, and below, where the pharynx ends and esophagus begins. Each of these intervals—above, between, and below the free borders of constrictor muscles—admits one or more structures into the pharynx: the cartilaginous portion of the auditory tube, levator palati muscle, and the ascending palatine branch of the facial artery enter above the superior constrictor by penetrating the fascia of interval (1); stylopharyngeus muscle and glossopharyngeal nerve (N. IX) gain access via interval (2); the internal branch of the superior laryngeal nerve and the superior laryngeal branch of the superior thyroid artery reach the interior of the pharynx through interval (3); and the recurrent laryngeal nerve and the inferior laryngeal branch of the inferior thyroid artery pass below the inferior constrictor muscle, interval (4).

The pharyngeal constrictors, though striated muscle, are not under voluntary control. They are innervated by special visceral motor fibers of the pharyngeal plexus derived from NN. IX and X–XI, *except* that the inferior constrictors (and the cricothyroid muscles) are supplied directly by the external branch of the superior laryngeal nerve.

Buccopharyngeal Fascia

The outermost coat of the pharynx, its serosa, is a thin, insubstantial layer of connective tissue given the proper name of **buccopharyngeal fascia**. It forms the anterior wall of a tissue space, the **retropharyngeal space**, that separates the pharynx in front from the vertebral column and prevertebral muscle and fascia behind. Its presence enables the pharynx to move on the vertebral column and to expand when a bolus of food passes through it. At the same time, however, the retropharyngeal space is one route by which infections may spread from side to side, and from the neck into the posterior mediastinum.

The Pharyngeal Lymphoid Ring

Also known as Waldeyer's ring, it is a collection of tonsils encircling the entrance into the pharynx (Fig. 15.6). Waldeyer's ring is said to be a barrier against infectious agents or other potentially harmful environmental hazards.

The **pharyngeal tonsil** is an accumulation of lymphoid tissue in the roof of the pharynx. It is small or absent in the adult, but prominent in children. When abnormally large, the "adenoids," as the pharyngeal tonsil is then called, is likely to obstruct nasal breathing. It was common practice to remove the lymphoid mass surgically so as to prevent children from becoming chronic mouth breathers.

The **tubal tonsils** lie near the pharyngeal openings of the auditory tubes and, thus, are located inferolaterally to the pharyngeal tonsil. When enlarged, the tubal tonsils may interfere with the equalization of atmospheric pressure on the two sides of the tympanic membrane.

The **palatine tonsils**, *the* tonsils, measure approximately two centimeters along their oval axes. Each occupies the triangular sinus between the palatine arches of the oropharyngeal isthmus and is attached by its fibrous capsule to the pharyngobasilar fascia. An immediate lateral relation of the tonsil is the facial artery from which it is separated only by the fascia and the superior constrictor muscle! Both ascending palatine and tonsillar branches of the facial artery supply the tonsil, as do palatine branches of the ascending pharyngeal artery, dorsal lingual branches of the lingual artery, and lesser (descending) palatine branches of the maxillary artery. The palatine tonsils, therefore, also are strategically important as sites where vessels of the external carotid artery anastomose freely. Given the highly vascular nature of the tonsils, their removal, tonsillectomy, may cause significant blood loss. The procedure requires considerable skill on the part of the surgeon and thorough familiarity with the anatomy of the region.

The nodular lymphoid tissue present in the root of the tongue makes up the **lingual tonsil**. It is situated inferomedially to the palatine tonsils, and forms the floor of Waldeyer's ring.

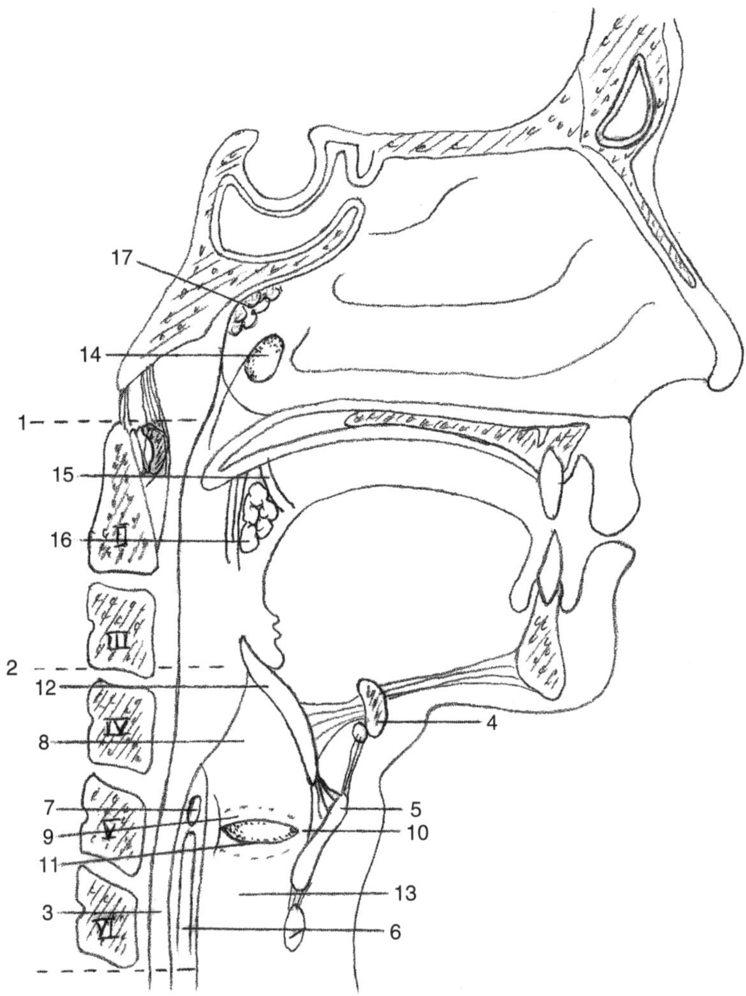

1. End of nasopharynx
2. End of oropharynx
3. Laryngopharynx
4. Hyoid bone
5. Thyroid cartilage (Adam's apple)
6. Arch and lamina of cricoid cartilage
7. Arytenoid cartilage
8. Vestibule and quadrangular membrane
9. Vestibular fold
10. Glottis
11. Vocal fold
12. Epiglottis
13. Infraglottic cavity
14. Torus tubarius
15. Anterior pillar of fauces
16. Palatine tonsil
17. Pharyngeal tonsil

15.1 Pharynx and larynx

Figure 15.1: Divisions of the pharynx

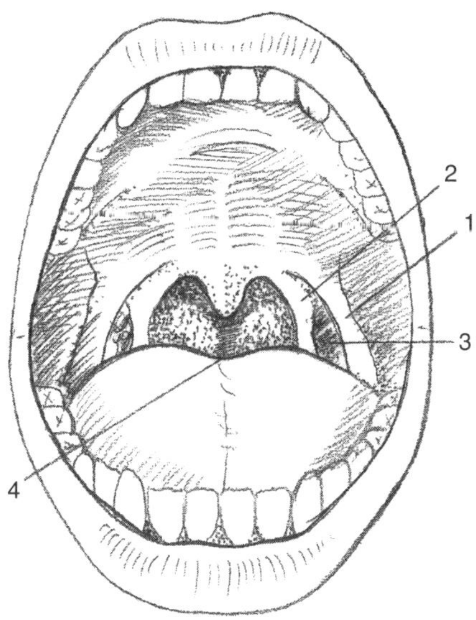

15.2 Oropharyngeal isthmus

1. Anterior pillar of fauces
2. Posterior pillar of fauces
3. Tonsillar sinus
4. Sulcus terminalis of tongue

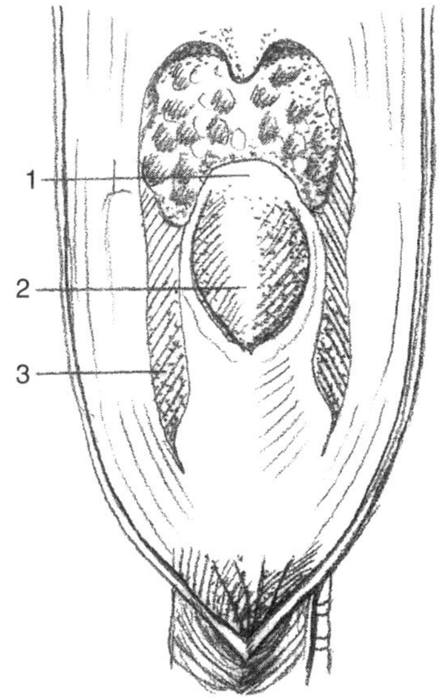

15.3 Laryngopharynx (posterior view)

1. Epiglottis
2. Laryngeal inlet
3. Piriform fossa

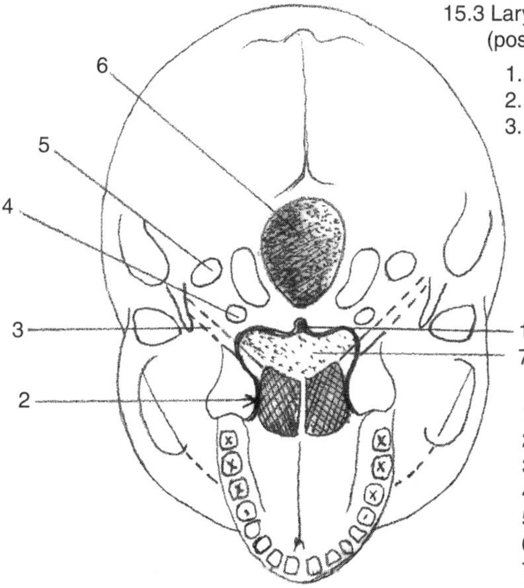

1. Pharyngeal tubercle
2. Medial pterygoid plate
3. Auditory tube
4. Carotid canal
5. Jugular foramen
6. Foramen magnum
7. Nasopharynx

15.4 Pharyngobasilar fascia attaching to skull

Figure 15.2: Oropharyngeal isthmus
Figure 15.3: Laryngopharynx
Figure 15.4: Pharyngobasilar fascia attaching to the skull

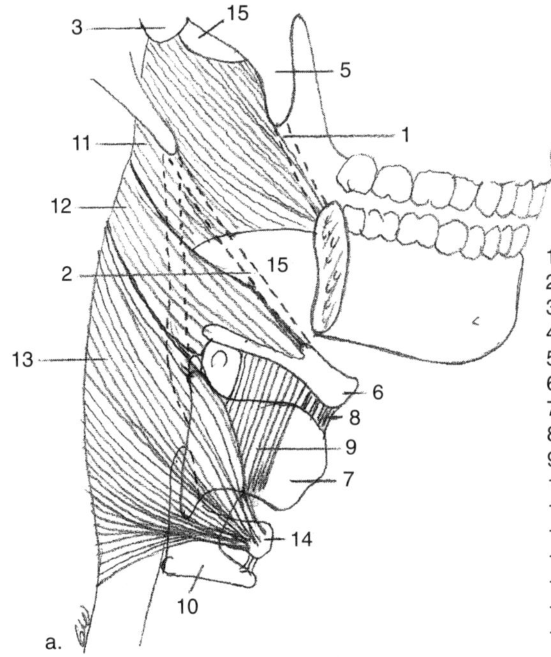

1. Sphenomandibular ligament
2. Stylohyoid ligament
3. Pharyngeal tubercle
4. Pharyngeal raphe
5. Medial pterygoid plate
6. Hyoid bone
7. Thyroid cartilage
8. Thyrohyoid membrane
9. Thyrohyoid muscle
10. Cricoid cartilage
11. Superior constrictor muscle
12. Middle constrictor muscle
13. Inferior constrictor muscle
14. Cricothyroid muscle
15. Pharyngobasilar fascia
16. Esophagus

15.5a-b Constrictor muscles a. Lateral view b. Posterior view

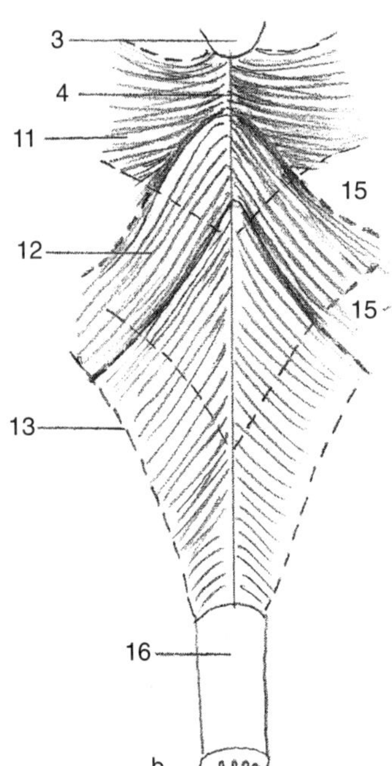

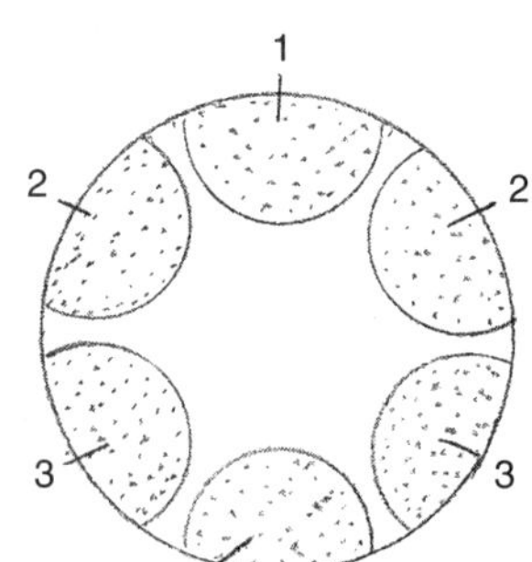

15.6 Pharyngeal lymphoid ring

1. Pharyngeal tonsil
2. Tubal tonsil
3. Palatine tonsil
4. Lingual tonsil

Figure 15.5a: The constrictor muscles a-lateral view
Figure 15.5b: The constrictor muscles b-posterior view
Figure 15.6: Pharyngeal lymphoid ring

Chapter 16
THE LARYNX

The larynx is a fibrocartilaginous organ. It is connected by a membrane to the hyoid bone, floating high above in the neck region. And, therefore, as the hyoid moves up or down, sideways, forwards or backwards, so does the larynx. It rises during deglutition and vocalization, as you can verify by observation and palpation.

The larynx is separated from the vertebral column behind by the laryngopharyngeal wall, retropharyngeal space, prevertebral muscles and fascia. It begins opposite the disc between vertebrae C3–4, and ends opposite vertebra C6 to become continuous with the trachea. Anterior relations are the infrahyoid muscles and fascia, upper thyroid gland, and, more laterally, the carotid sheath and contents (see Fig. 12.6, p. 83).

The larynx is a hollow organ, divided, not unlike the pharynx, into three parts: vestibule, glottis, and infraglottic cavity.

The Division of the Larynx

The **vestibule** extends from the laryngeal inlet above to the vestibular ligaments below. The epiglottis, a spoon-shaped piece of cartilage, guards the entrance into the vestibule. The vestibular ligaments are the foundation of the **vestibular folds** (false vocal cords).

The **glottis** of the larynx is the space between the vestibular ligaments above and the vocal ligaments below. The latter form the framework of the **vocal folds**, (true vocal cords). The glottis communicates laterally with the **laryngeal saccule**—a pouch of mucous membrane that in great apes extends under the armpit. By pumping their arms, bellows-like, apes are able to generate ear-splitting sounds.

The **infraglottis cavity** is the part of the larynx below the vocal ligaments. It starts at the level of the rima glottidis, the wedge-shaped interval between the vocal folds, and ends where the trachea begins.

The Cartilages of the Larynx

The **thyroid cartilage**, which is connected above to the hyoid bone by means of the thyrohyoid membrane, is composed of two laminae (plates) united anteriorly to form the **laryngeal prominence** or "Adam's apple" (Figs. 16.1 and 15.1, p. 104). It is more pronounced in men, because the two laminae meet at a 90° angle, whereas in women the angle is a shallow 120°. This sexual dimorphism should remind you of similar sex differences at the subpubic angle.

The **cricoid cartilage** is commonly compared to a signet ring, the narrow arch facing front and the broad lamina pointing in the opposite direction. Thyroid and cricoid cartilages are linked behind by means of synovial joints, which enable the thyroid cartilage to tilt forwards on the cricoid. The lower margin of the cricoid cartilage, opposite vertebra C6, is connected to the trachea by the cricotracheal membrane.

Two **arytenoid cartilages** are mounted atop the cricoid lamina on either side of the median plane. Each is shaped like a three-sided pyramid having medial, anterolateral, and posterior surfaces and anterior, posterior, and lateral borders. The anterior border projects forwards as the **vocal process** and the lateral border is prolonged as the **muscular process**. The base of each arytenoid cartilage, and the area of cricoid lamina in contact with it, form the crico-aryteroid joint. The apex of each arytenoid cartilage supports the small, and quite insignificant, elastic corniculate and cuneiform cartilages. The arytenoids, likewise, are composed substantially of elastic cartilage; only the base articulating with the cricoid is, in typical fashion, hyaline cartilage.

The **epiglottis**, another piece of elastic cartilage, is attached below by its apex to the deep aspect of the laryngeal prominence. Capable of folding over the laryngeal inlet, the epiglottis is one of several devices preventing solids and liquids from entering the airway.

The Membranes of the Larynx

The **thyrohyoid membrane**, between the hyoid bone and the upper margin of the thyroid cartilage, transmits the pull of muscles that move the hyoid bone (Fig. 16.1). It is thick in front, where it forms the median thyrohyoid ligament; its free posterior borders are the lateral thyrohyoid ligaments, round elastic folds within each of which is a triticeal cartilage. The thyrohyoid membrane is pierced by internal (sensory) branches of superior laryngeal nerves and companion vessels distributing to the tissues of the vestibule.

The quadrangular membrane lines the inner aspect of the epiglottis and the vestibular part of the larynx (Fig. 15.1, p. 104). Its free upper border stretches between the lower one-half of the epiglottis and the tips of the arytenoid cartilages. As **ary-epiglottic folds** and **muscles**, the upper border of the quadrangular membrane forms the sides of the laryngeal inlet. The free lower border of the membrane forms the vestibular folds.

The **cricovocal membrane** also has upper and lower borders (Fig. 16.1, p. 104). The lower border is firmly attached to the upper margin of the cricoid cartilage. Its upper border, however, is free—except at its extreme ends: anteriorly, the membrane inserts into the deep aspect of the laryngeal prominence as the median cricothyroid ligament; posteriorly, it connects to the vocal processes of the arytenoid cartilages. The right and left free upper borders of the membrane created by this three-pronged attachment are the vocal ligaments, the ligamentous foundations of the vocal folds. The narrow triangular gap between them is the **rima glottidis** (Fig. 16.2). The dome-shaped roof formed over the infraglottic cavity by the two sides of the cricovocal membrane is known as *conus elasticus*. The choice of words here serves as a helpful reminder that the cricovocal membrane is highly elastic, thus able to rebound after being forcibly displaced by the passage of exhaled air. It is this property that is crucial to the functional capacities of the larynx, as will become clear in a moment.

The Muscles of the Larynx

The larynx as a whole is elevated by muscles acting on the hyoid bone from above; thus, contraction of mylohyoid (innervated by N. V) flattens the angle of the muscle at its raphe and hauls the larynx upwards. Other suprahyoid muscles, such as the geniohyoids (n. C1) and stylohyoids (N. VII), even digastrics (N. V and VII), also are elevators of the larynx when contracting in unison. Acting together, paired geniohyoid muscles protract the larynx and stylohyoids retract it. The larynx is shifted to the right or left when geniohyoid and stylohyoid muscles of the same side contract simultaneously. The stylopharyngeus muscle (N. IX) attaches to the larynx directly and, by rights, should be called stylopharyngeus. The same case may be made for salpingopharyngeus and palatopharyngeus muscles, because their primary attachment also is to the posterior border and/or the superior horn of the thyroid cartilage.

Gravity lowers the larynx when suprahyoid muscles relax, but infrahyoid muscles exist to depress it. Their main function, however, may be to exercise control over the elevators of the larynx by "putting on the brakes" or "paying out rope." Known also as "strap muscles," the depressor hyoid muscles act on the hyoid bone from below. Only the sternothyroid muscles attach to the larynx directly, and they depress the hyoid bone by way of the thyrohyoid membrane.

The functional capacities of the larynx depend on the actions of muscles attaching to the muscular processes of the arytenoid cartilages and freely movable crico-arytenoid joints. They allow for (1) medial and lateral rotation of arytenoid cartilages around their long axes, and (2) their physical separation and approximation in the frontal plane.

Muscles attaching to the muscular process bring about changes in the positions of arytenoid cartilages and vocal folds (Fig 16.3). Paired **posterior**

crico-arytenoid muscles abduct the vocal folds, thus widening the rima glottidis, by rotating the arytenoids laterally about their long axis and/or moving them apart bodily—like sliding doors—up or down the sloping shoulders of the cricoid cartilage (Fig 16.3). The rotary motion achieved by horizontal fibers of the muscle is countered by the pull of the **lateral crico-arytenoids muscles**, and the translatory effect of the oblique fibers of the posterior crico-arytenoids is opposed by the action of the (unpaired) **transverse arytenoid muscle**, which connects the posterior surfaces of the two arytenoid cartilages (Fig. 16.3). Lateral crico-arytenoid and transverse arytenoid muscles, therefore, are adductors of the vocal folds; they narrow the rima glottidis and, indeed, may close down the airway completely.

Lengthening and shortening of the vocal folds is accomplished also by functionally opposing pairs of muscles. The **cricothyroid muscles** lengthen the folds as they tilt the thyroid cartilage towards the cricoid arch (Fig. 16.4). Their opponents are the **thyro-arytenoid muscles**, which lie embedded in the cricovocal membrane. Tensing of the vocal cords, as distinguished from lenghtening, results when **vocalis** contracts isometrically. The vertical depth of the folds is increased at the same time.

The cricothyroid muscles, like the inferior pharyngeal constrictors, are innervated by the external branch of the superior laryngeal nerve. All other intrinsic muscles of the larynx are supplied by the recurrent laryngeal nerve, which also is sensory to the mucous membranes below the vocal folds. Injury to the nerve generally produces abductor muscle paralysis, thus narrowing of the rima glottidis. Unilateral nerve damage results in high-pitched, noisy respiration—respiratory stridor; paralysis of both posterior crico-arytenoid muscles may lead to death by suffocation.

Functions of the Larynx

Being part of the upper respiratory tract, the essential function of the larynx is to prevent solids, liquids, and foreign bodies from entering the trachea. Towards that end, the vocal folds become fully adducted during swallowing (while breathing stops), ary-epiglottic muscles contract causing the epiglottis to fold over the laryngeal inlet, and the larynx as a whole is elevated and tucked under the root of the tongue.

Closure of the laryngeal inlet also closes off the thoracic cage. "Holding one's breath" fixes the ribs in inspiration and aids in the lifting of heavy objects. It also increases intra-abdominal pressure, which allows "bearing down" during urination, defecation, and child birth.

Vocalization, sound production, is a non-specific, incidental function of the larynx; indeed, the vocal organ of birds, the syrinx, is altogether independent of the larynx. Vocalization normally takes place in the expiratory phase of respiration. The act requires that air be expelled from the lungs while the vocal folds are fully apposed. Then, whenever the force of exhaled air exceeds the elastic tension of the folds, they are blown apart. Remember that the vocal folds bear no resemblance to musical strings that are set in motion when struck or plucked. The vocal folds are, quite simply, the free upper borders of the cricovocal membrane; and, when a current of air is blown past them, the folds vibrate in the manner of a reed, shaping the emerging stream into wave forms of certain frequency. Vibration frequency depends on the force generated by changes in intrapulmonic pressure, air speed, and upon length, tension and mass of the vocal folds. These factors together determine the *pitch* of sound; *i.e.*, high and low notes on a scale. The *quality* of sound depends on the size and shape of the larynx, pharynx, mouth, and nose—including paranasal sinuses. *Articulation* is influenced by the position of the tongue and lips, and by dentition.

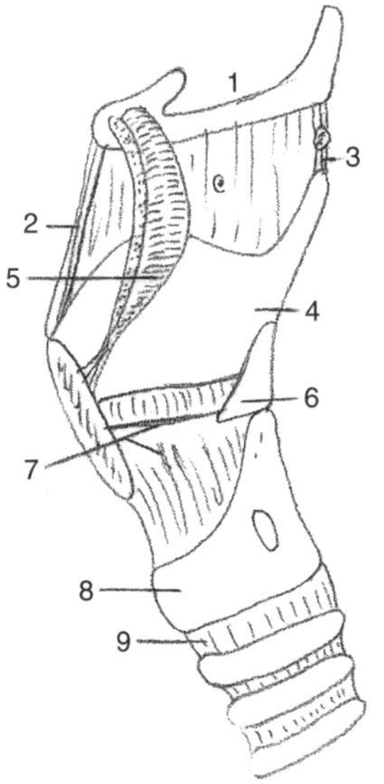

1. Right half of hyoid bone
2. Median thyrohyoid ligament
3. Lateral thyrohyoid ligament
4. Right half of thyroid cartilage
5. Epiglottis
6. Vocal process of left arytenoid cartilage
7. Vocal fold and cricovocal membrane
8. Cricoid cartilage
9. Cricotracheal ligament

16.1 Cartilages and membranes of larynx

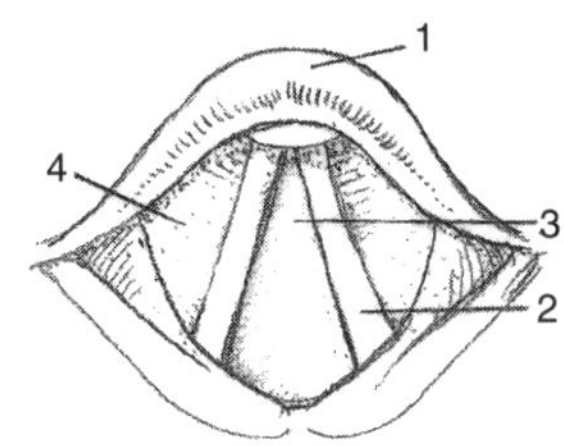

16.2 Rima glottidis (superior view)

1. Epiglottis 3. Rima glottidis
2. Vocal fold 4. Cricovocal membrane

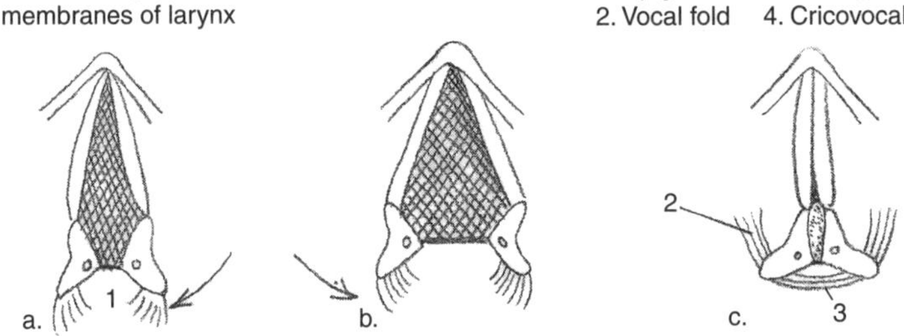

16.3 a-c Muscles acting on vocal folds

1. Posterior crico-arytenoids abduct folds (a), (b)
2. Lateral crico-arytenoids and transverse arytenoid adduct folds (c)
3. Transverse arytenoid

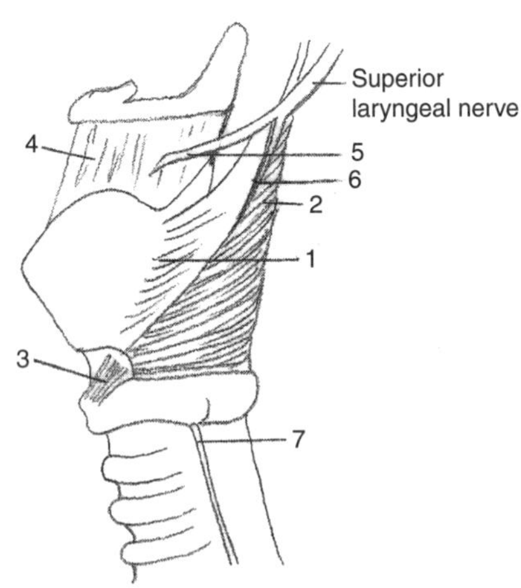

16.4 Cricothyroid muscle and inferior constrictor muscle

1. Oblique line of thyroid cartilage
2. Inferior constrictor muscle
3. Cricothyroid muscle
4. Thyrohyoid membrane
5. Internal branch
6. External branch
7. Recurrent laryngeal nerve

Figure 16.1: The cartilages and membranes of the larynx
Figure 16.2: The rima glottidis
Figure 16.3: a-c Action of muscles on arytenoid cartilages
Figure 16.4: Cricothyroid muscle and inferior constrictor muscle

Chapter 17
INTRODUCTION TO THE LYMPHATIC SYSTEM

The lymphatic system is a collection of vessels, nodes, and organs including the pharyngeal lymphoid ring. Lymph, the fluid flowing through the system, is colorless (L., *lympha* = clear water) and the vessels normally containing it are usually collapsed and difficult to isolate. Hence, the lymphatic system is frequently perceived as being less important than other systems of the body and, in general, given short shrift in textbooks of topographic anatomy. Actually, of course, the lymphatic system is a vital cog in the homeostatic machinery of the body. One of its functions is to return to the bloodvascular system fluids and plasma proteins that were filtered or forced out at the arterial end of capillaries and lost to tissue spaces. By returning extravasated plasma proteins, especially albumins, to the blood stream, the lymphatic system helps maintain plasma osmotic pressure and, thus, the fluid balance between blood and the interstitial spaces.

The lymphatic system has additional important functions in line with its duties as part of the body's defense establishment. It is responsible for the proliferation and dissemination of lymphocytes, assumes a role in the regulation of the immune response, and participates in the removal of foreign particulate matter. We have already discussed the significance of lymph nodes as filters that trap and neutralize potentially harmful material.

General Plan

Lymph vessels are found almost everywhere in the body, and they come in all sizes. The lymphatic system lacks a muscular pump or heart, and lymph is moved through the sytem of tubes by the milking and massaging action of pulsating arteries with which lymphatics travel, and of contracting muscles with which they are in contact.

Lymphatic capillaries permeate subepithelial tissues. They are especially abundant at mucocutaneous junctions; *i.e.*, eyelids, external nares, lips, anus, and vagina, and absent from avascular tissue (*e.g.*, epidermis and cartilage) and avascular organs (*e.g.*, cornea and lens of the eye). They appear to be lacking in the alveoli of the lung, in bone marrow, the spleen, the placenta, and in the central nervous system where tissue fluid flows in perivascular sheaths.

In the skin, lymphatic capillaries form a superficial network of closed loops which communicates with a deep capillary plexus. The lymphatics of the deep plexus increase in size and become collecting vessels which conduct lymph towards lymph nodes. They also acquire one-way valves that, when closely spaced, confer on distended vessels a beaded appearance. Efferent lymphatics from groups of nodes form trunks of increasing sizes; those draining the right and left nodes of the abdominopelvic region form the **right** and **left lumbar trunks** whose confluence, just below the aortic hiatus, marks the beginning of the **thoracic duct** (Fig. 17.1). An unpaired **intestinal trunk**, carrying most of the absorbed fats from the intestine in the form of chylomicrons, may form a third limb—in which case the root of the thoracic duct is expanded and termed the **Cisterna chyli**. The paired (descending) **intercostal trunks**, draining the lower intercostal spaces, terminate in the thoracic duct as well.

As you are well aware, the thoracic duct pours its contents into the cardiovascular system where left jugular and subclavian veins meet. Left **subclavian, jugular**, and **bronchomediastinal lymph trunks** usually join the thoracic duct before it ends. Lymph from the right hemi-thorax, the right upper limb, and the right side of head and neck is brought to the right jugulosubclavian angle; the right subclavian, jugular, and bronchiomediastinal trunks frequently unite to form a **right lymphatic duct**.

Lymph nodes are inserted into the system at strategic points, and most of the lymph passes through at least one group of nodes, and generally more. Once lymph has entered collecting vessels possessing valves its subsequent course is fixed. Up to that point, however, lymph seeks the path of least resistance—which accounts for variations in textbook descriptions of regional lymphatic drainage.

Lymph nodes are numerous in the axillary and inguinal regions and throughout the head and neck. When abnormally large or swollen, such superficial nodes can be palpated with relative ease. There are, of course, many more deep lymph nodes in the trunk than exist in the limbs.

In general, collecting vessels accompany arteries in their passage through the body, and lymph carried within, therefore, travels a fairly direct and predictable course. All bets are off, however, when obstruction resulting from pathology or disease interferes with normal patterns of lymphatic drainage in the system.

In both upper and lower limbs, superficial lymphatic vessels run alongside superficial veins, *e.g.*, cephalic and saphenous veins, and lymph leaving the superficial nodes of a limb usually flows towards its deep nodes.

Clinical Considerations

Lymph vessels are channels through which bacteria or other pathogens may be carried far afield from the original site of infection. Thus, an abcess at the tip of a finger may develop into an infection of lymphatic vessels of the arm (lymphangitis) and of the lymph nodes in the axilla (lymphadenitis). Such disease may terminate in blood-poisoning (septicemia) and death. Prolonged obstruction of lymph flow is associated with the condition of lymphedema, the accumulation of interstitial fluid in tissue spaces.

Lymph vessels are pathways also for the spread of neoplastic disease. Carcinomas and adenomas, *i.e.*, cancers that originate in epithelial organs, tend to invade near-by lymphatic capillaries at an early stage and to metastasize rapidly from the primary site to other parts of the body. Familiarity with fields of lymphatic drainage and locations of regional lymph nodes is important, for in the surgical treatment of malignant disease one strategy calls for the stripping of lymph nodes and vessels downstream from the lesion. And, once malignancy cells have reached lymph vessels that are inaccessible to the surgeon, cancers become inoperable.

Many important structures situated in close proximity to infected or inflamed lymph nodes may become secondarily involved, the effects of which can be far-reaching. Swollen tracheobronchial nodes, for example, pressing on the (left) recurrent laryngeal nerve, may cause speech disturbances; and inflammation of occipital lymph nodes is likely to produce neuritis of the greater occipital nerve. Other neuroinflammatory reactions may be expected where axillary nodes surround the intercostobrachial nerve, where parotid nodes lie on N. VII, where the phrenic nerve makes contact with supraclavicular nodes, and where the spinal accessory nerve crosses the posterior triangle (Fig. 17.2).

Lymphatic Drainage of Body Wall

Lymph drainage from the body's surface is in the watershed mode; *i.e.*, lymph flows away in opposite directions from horizontal and vertical axes intersecting at the level of the umbilicus. It converges upon ipsilateral axillary and superficial inguinal lymph nodes, though lymph near the intersecting lines may cross the territorial divide. **Axillary nodes**, therefore, receive lymph from the body's upper surface. That originating in front drains to pectoral members of the group; and lymph from the subcutaneous tissue of the upper back, including that of the breast's axillary tail, is brought to the subscapular group of axillary nodes.

Superficial inguinal lymph nodes reveive lymph from the lower body wall. Some members of this group (inferior nodes) are arranged in a vertical chain on the lateral side of the great saphenous vein; the rest of the superficial inguinal nodes forms a horizontal row below the inguinal ligament, both lateral (superolateral nodes) and medial (superomedial nodes) to the vertical chain. Subcutaneous lymph from the lower back goes to the superolateral nodes, and that originating from in front below the waist ends up in the superomedial nodes. They also receive lymph from the perineum (urethra, anus, external genitalia), but not from the testes or the ovaries. Lymph that arises in either of these organs follows the path of the gonadal arteries to **abdominopelvic nodes**.

Lymphatic Drainage of the Breast Revisited

The critically important topic of lymph flow in breast tissue was discussed in connection with the anatomy of the pectoral region (chapter 7). This may be the place to revisit certain aspects of the subject, and to conclude that the watershed model just posited for the lymphatic drainage of the body's surface does not apply to the breast. It has been estimated that ten percent of all women having gone through mastectomy once develop cancer in the other breast—presumably, because lymph can cross to the opposite side when malignancy blocks ipsilateral drainage.

Breast lymph of the outer quadrants drains alongside lateral thoracic and subscapular arteries (Fig. 17.3). Most of it is channeled to axillary nodes (pectoral and subscapular), whose efferent vessels terminate in the subclavian lymph trunk. Breast lymph of inner quadrants (ca. 25 percent of total) follows the internal mammary branches of the internal thoracic artery to parsternal nodes. This lymph also may end up in the subclavian trunk—though, it is just as likely to pass alongside the internal thoracic arteries into the neck. After filtering through deep cervical lymph nodes, the lymph will reach the jugular lymph trunk. The same may be said for lymph travelling with branches of the thoraco-acromial artery to the infraclavicular nodes of the deltopectoral triangle (deltopectoral group) and the supraclavicular nodes of the posterior triangle.

Lymphatic Drainage of Head and Neck

All lymph of head and neck is ultimately brought to the deep cervical lymph nodes, whose efferent vessels are tributaries of the jugular lymph trunk. The **deep cervical nodes** form vertical chains around the internal jugular veins, but, before reaching this group of nodes, most lymph has already filtered through one or more outlying stations of nodes. Some of these are deeply placed and receive lymph from the thyroid gland and from upper respiratory and digestive passages; nodes of this internal "necklace" include **pre-** and **paratracheal** as well as **retropharyngeal nodes**. Other collections of lymph nodes are related to both anterior and external jugular veins and their tributaries; they form an external "necklace" and filter lymph from the superficial tissues of head and neck. The **submental nodes**, for example, intercept lymph from the floor of the mouth, from the lower gums and lips opposite the lower four incisor teeth, and from the tip of the tongue (Fig. 17.2). Lymph from the rest of the mouth and tongue passes to nodes present in the submandibular triangle. The **submandibular nodes** are recipients also of lymph from a triangular area encompassing the center of the forehead, the nose, and adjacent cheeks. Other members of the external necklace are the **facial nodes**, which receive lymph from forehead, temples, orbits, and eyelids; the **parotid nodes**, draining cheeks and external auditory meati; and the **mastoid** and **occipital nodes**, which filter lymph originating from the auricles and posterior scalp.

Two lymph nodes of the deep cervical chain deserve name recognition. Represented bilaterally, they are located where the internal jugular vein is crossed by muscle of the anterior triangle. The **jugulodigastric nodes**, in the path of effluent lymph from the palatine tonsils and surrounding tissue, react to tonsillitis and sore throat. The **jugulo-omohyoid nodes**, likewise, are clinically important, because they receive lymph from all parts of the tongue and become involved in carcinoma of the organ.

Lymphatic Drainage of the Tongue

Tongue-related lymph has no well-defined watershed flow, though its drainage does seem to obey certain laws; *e.g.*, the farther forward lymphatic vessels arise on the tongue, the lower down they terminate on the chain of deep cervical nodes. Be aware that lymph from the tip of the tongue may cross the center line and, thus, reach both ipsi- and contralateral nodes. Lymph from the rest of the anterior two-thirds of the tongue drains ipsilaterally—though there is some contralateral flow near the midline. Lymph from the posterior one-third of the tongue may go either way.

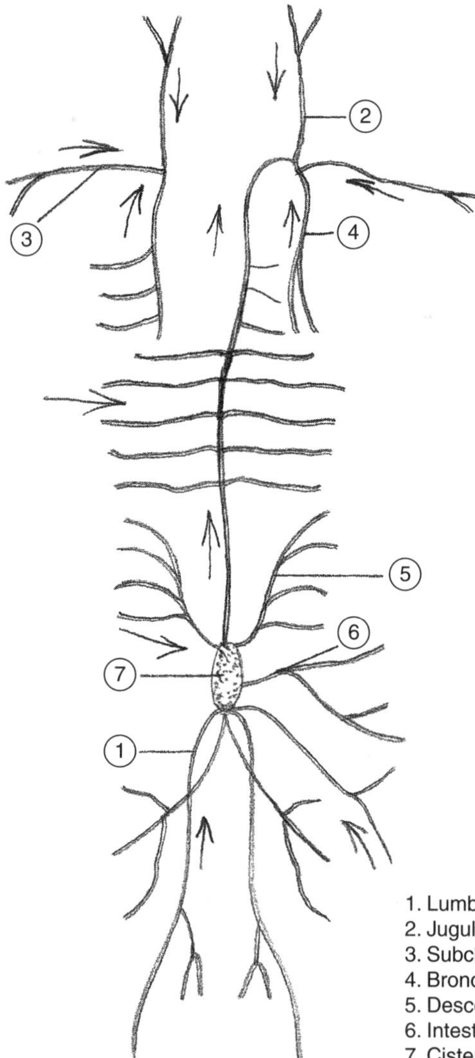

1. Lumbar trunk
2. Jugular trunk
3. Subclavian trunk
4. Bronchomediastinal trunk
5. Descending intercostal trunk
6. Intestinal trunk
7. Cisterna chyli

Figure 17.1: Thoracic duct and tributaries

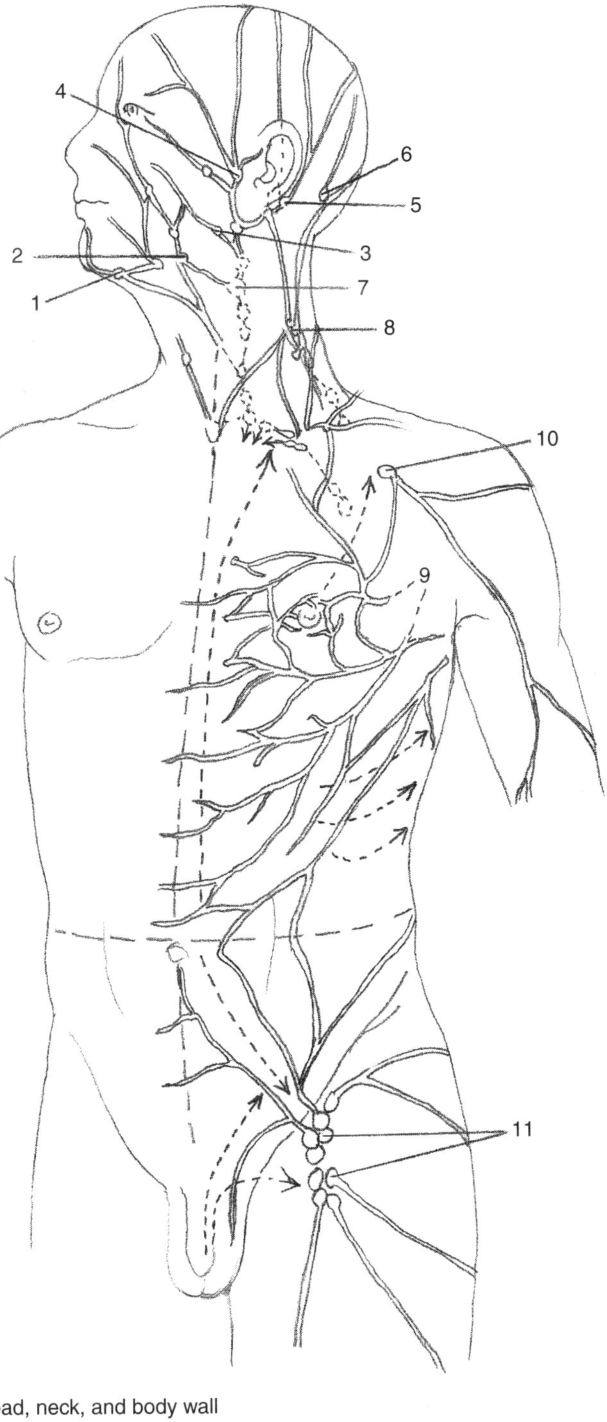

1. Submental node
2. Submandibular node
3. Facial node
4. Parotid node
5. Mastoid node
6. Occipital node
7. Deep cervical node
8. Node of posterior triangle
9. Location of axillary nodes
10. Infraclavicular node
11. Superficial inguinal nodes (inferior, superolateral, and superomedial)

17.2 Lymphatic drainage of head, neck, and body wall

Figure 17.2: Lymphatic drainage of head and neck
Figure 17.2: Lymphatic drainage of the body wall

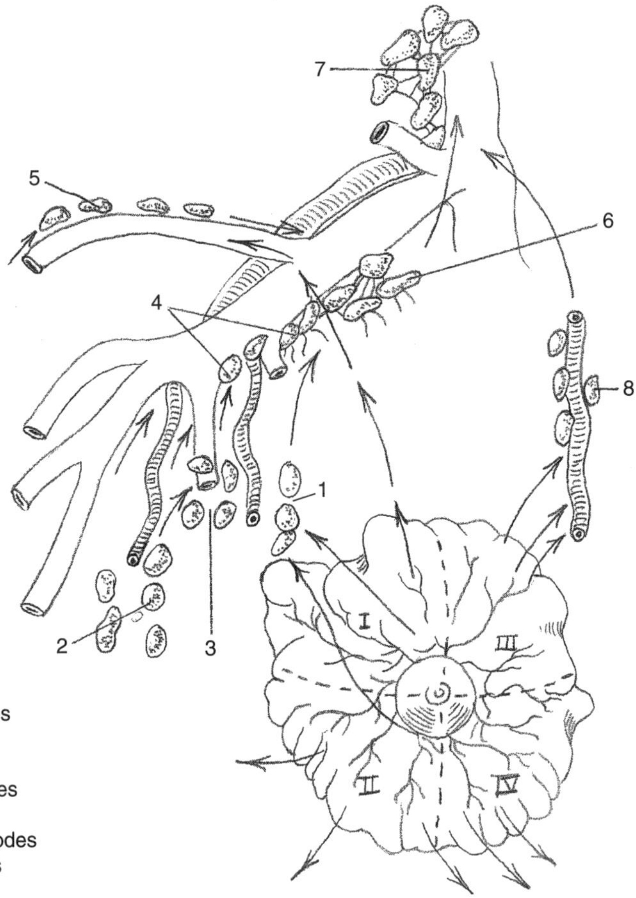

1. Pectoral nodes
2. Subscapular nodes
3. Central nodes
4. Lateral nodes
5. Deltopectoral nodes
6. Apical nodes
7. Supraclavicular nodes
8. Parasternal nodes

17.3 Lymphatic drainage of breast

Figure 17.3: Lymphatic drainage of breast

Chapter 18
ARTHROLOGY

Arthrology is the study of joints. It, perhaps, more than any other subject of anatomy illustrates the close relationship between structure and function, for arthrology is the study of how bones are joined to permit (or prevent) movement. Joints, therefore, are crucial components of the human frame. Indeed, they play such a pivotal role in musculoskeletal anatomy that, rather than thinking of the human skeleton as an assembly of bones, you should look upon it as a system of joints connected by lengths of levers.

The joints of the body may be classified functionally as either *freely movable, slightly movable,* or *immovable*. And yet, you should be aware that joints are subject to change in the lifetime of an individual. Some undergo structural modification shortly after birth or during adolescence as part of normal and natural postnatal skeletal development, growth, and maturation. Accordingly, certain bony unions will permit movement early in life, but not later. Likewise, many movable joints undergo degenerative change with age; they may be rendered immovable by genetic predisposition, normal wear and tear, or disease.

Joints vary in structure and, therefore, differ in their functional capacities. They are numerous in locations that require flexibility and mobility—*e.g.*, in limbs, which are adapted for prehensile and locomotor functions; or the thorax, whose distensible walls are the engine that drives respiration. Conversely, parts of the body that require stability, security, and permanence have fewer joints, and they are quite differently structured than those engaged in movement. The head, for instance, has only one freely movable joint (if you exclude the chain of ear ossicles); namely, the temporomandibular joint.

Classification of Joints

The joints of the body may be classified according to functional capacity or structural design, but a broad typology that incorporates both properties is clearly most useful. For, after all, whether a joint permits movement or not depends surely on the medium—the glue—present at the site of bony contact. Some bones are literally cemented together—as are, for example, the cranial bones at their suture lines. At the other extreme are bones whose articular surfaces are physically separated by a fluid-filled space. Joints of this latter class are usually highly mobile; they are also most apt to suffer dislocation. In general, as previously discussed (see chapter 2), joint stability and mobility are inversely related. In freely movable joints, contact-area dimensions and bone-to-bone congruence are major factors ensuring their integrity, When these attributes are lacking, ligaments and muscles form first and second lines of defense against partial or complete joint separation.

Freely Movable Joints

Joints designed to permit a wide range of motions are classified as **diarthroses** (Gr., *di-* = apart), or **synovial joints**. All have certain characteristics in common (see Fig. 6.4, p. 40): (1) articular cartilage, which coats the epiphysial ends of apposing bones, (2) a joint cavity lined by synovial membrane, and (3) a sturdy joint capsule.

The cartilage covering the articular surfaces of bones differs from hyaline cartilage elsewhere, *e.g.*, costal cartilage, by lacking a surface membrane (perichondrium). But, like all cartilage, articular cartilage is avascular and has little regenerative capacity; hence, when it is eroded—owing to normal wear and tear, or as the result of traumatic injury—the underlying bone lies exposed. Bony outcroppings, exostoses, will then develop—producing spurs or lipping that may limit movement and cause pain.

The joint cavity contains a colorless, viscous liquid, synovial fluid, which is believed to originate as a filtrate of blood plasma from the rich capillary plexus in the synovial membrane. The viscocity is due to the presence of hyaluronic acid, a mucopolysaccharide, produced by one cell type of the membrane surface layer. A second cell type appears to participate in the removal of detritus by phagocytosis and pinocytosis. The quantity of synovial fluid varies from joint to joint. The knee joint, the largest synovial joint in the body, contains probably less than one milliliter of fluid; *i.e.*, the joint cavity normally is little more than a potential space. Inflammation of the synovial membrane, synovitis, or other joint diseases, causes fluid to accumulate with resulting discomfort and range of motion deficits.

The synovial membrane (synovium) is applied to the internal surface of a densely fibrous capsule that encloses the joint. The joint capsule, though frequently lax to facilitate unimpeded movement, is strengthened by ligaments that blend with it. Accessory ligaments, either external or internal to the capsule, further support the joint; and, both capsular and accessory ligaments are commonly oriented in such a manner as to resist joint separation forces. The capsule may be reinforced further by slips from tendons crossing the joint.

Types of Synovial Joints

Many structural varieties of synovial joint exist in the human body, some so constructed as to allow movement in only one plane while others permit motion in more than one direction.

Uniaxial joints are limited to the performance of just one functional modality, *i.e.*, either flexion-extension, abduction-adduction, or medial-lateral rotation. The humero-ulnar articulation at the elbow, a **hinge** (ginglymus) **joint** permitting only flexion and extension, is of this type (Fig. 18.1 a). So is the atlanto-axial joint which, like other **pivot** (trochoid) joints, allows rotation in only a single plane (Fig. 18.1 b).

Biaxial joints afford movement in two directions, though not necessarily to the same degree. Thus, the radiocarpal joint at the wrist can be flexed and extended as well as abducted and adducted. Observe, however, that flexion-extension is freer than abduction-adduction, and that the joint is so designed as to favor flexion over extension and abduction over adduction. The radiocarpal joint is classified as a **condylar joint** (Fig. 18.1 c), a bony union resembling an egg in a spoon. **Saddle joints** (Fig. 18.1 d), such as the carpometacarpal joint of the thumb, are similar to condylar joints but permit also a slight degree of rotation.

Multi- or polyaxial joints allow movement in all directions. The glenohumeral articulation at the shoulder is an **enarthrosis** (Fig. 18.1 e), a synovial **joint** of the ball and socket variety, which enables the arm to be moved through an infinite number of axes. **Plane joints** (Fig. 18.1 f) are diarthroses that permit sliding or gliding of joint surfaces which, generally, are flat or only mildly concavoconvex. Examples of such bony unions are the vertebral arch joints and the acromioclavicular articulation.

In a small number of synovial joints, the joint cavity is either partially or completely divided by a plate, disc, or crescent (meniscus) of fibrocartilage. Such complex synovial joints are capable of more than one function; *i.e.*, they may act as a hinge joint in one compartment of the joint cavity and as a plane joint in the other. The temporomandibular joint, to be discussed at the end of this chapter, is one of several complex joints in the body.

Finally, I should note that not all synovial joints are necessarily freely movable. The sacro-iliac joints, for instance, though equipped with a small joint space containing synovial fluid, are quite immobile. Both are firmly secured by a strong capsule and an abundance of capsular and accessory ligaments. Thus, they are able to sustain a vertical load and transmit it to the lower extremities.

Slightly Movable Joints

Articulations that are structured to permit a slight amount of motion, but lack a joint cavity and other anatomical features of synovial joints, are defined as **amphiarthoses** (Gr., *amphi* = both, on both sides). Structurally, they are classified as either symphyses, syndesmoses, or synchondroses.

Symphyses (Gr., *Sym-physis* = grow together) are typically located in the median plane. The bones are united by fibrocartilage, as are the two hip bones of the symphysis pubis (Fig. 18.2). Intervertebral joints, *i.e.*, vertebral bodies and intervening intervertebral discs, belong in this category as does the symphysis menti (L., *mentum* = chin). Parenthetically speaking, the symphysis menti forms during the fourth week of embryonic development, when right and left mandibular processes meet in front to become lower jaw. At first a slightly movable joint, the symphysis menti persists in this form in some animals, *e.g.*, the dog. In the human, of course, the two mandibular processes fuse, thus converting the primitive symphysis menti into an immovable joint. A shallow midline depression on the chin postnatally marks the site of embryonic fusion which, if incomplete, leaves a pronounced dimple or cleft.

Joints consisting of bone united to bone by fibrous or membranous tissue are **syndesmoses** (Gr., *-desmo* = ligament), and articulations in which bony surfaces are joined by a bar or plate of hyaline cartilage are **synchondroses** (Gr., *-chondro* = cartilage). Both the ligamentous coracoclavicular connection and the intermediate radio-ulnar joint complete with interosseus membrane (Fig. 18.3) are syndesmoses, while the cartilaginous union of first rib and manubrium sterni is a synchondrosis.

Immovable Joints

Many amphiarthroses, not just the symphysis menti, are transitory. Present at birth, they are transformed postnatally into **synarthroses** (Gr., *syn* = union), *i.e.*, immovable joints, as membrane or cartilage is replaced by bone. Consider, for example, the **sutures**—immovable joints peculiar to the vault of the skull. At birth and for several months afterwards, the cranial bones are united by membrane to form the soft spots or fontanelles of the infant skull (Fig. 18.4). These membranous unions gradually disappear as cranial bones grow, make contact, and interlock at their margins. Joints whose bones articulate without intervening soft tissue are known as **synostoses** (Gr., *os* = bone).

You may also be aware that thin plates of hyaline cartilage separate epiphyses and diaphyses of long bones for many years after birth, and that the cartilage plates are sites where bones grow in length (Fig. 18.5). With onset of sexual maturity, when levels of sex hormones rise sharply, **closure of**

epiphyses occurs: bone replaces cartilage, the cartilaginous joint is transformed into a synostoses, and linear growth ceases.

Another synchondrosis that ossifies in the second decade of human life is present at the hip. Until then, the three flat bones of the os coxae are conjoined by a Y-shaped or triradiate piece of hyaline cartilage. It, too, gives way to bone, following which pubis, ilium, and ischium are solidly and permanently united.

The Temporomandibular Joint

The **temporomandibular** (T-M) **joint** is a synovial joint between the head of the mandible and the mandibular fossa on the temporal bone (Fig. 18.6). I have already pointed out that it is a complex synovial joint; *i.e.*, its joint cavity is divided by a disc of fibrocartilage, making it possible for two different movements to take place simultaneously. To prove the point, stick a finger in each ear. When you now open your mouth, you will feel disc and condylar process, moving as one, slide out of the mandibular fossa onto the articular eminence in a manner typical of a plane joint; while, concurrently, the condyle moves on the disc in the style of a hinge.

You should realize also that opening of the mouth, *i.e.*, depression of the mandible, especially against resistance, requires muscular force. The mandible is depressed when lateral pterygoid muscles draw the condyles forwards onto the articular eminence (protraction), at the same time that mylohyoid and digastric muscles pull the chin backwards. In the branch of physics known as mechanics, the concept of *force couples* refers to two equal but opposite forces not acting in the same line; it is invoked to explain disequilibrium with a net zero force. Think of the muscles pulling on the mandible in opposite directions as force couples whose horizontal pull opens the mouth by rotating the mandible about an axis passing through the mandibular foramina.

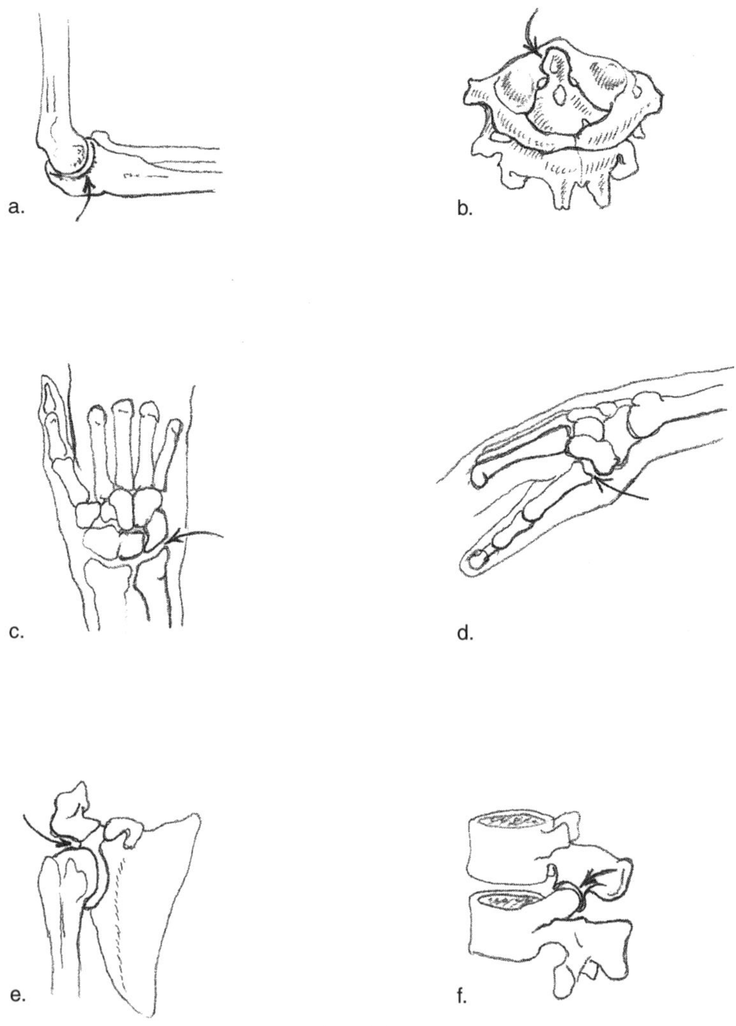

a. Hinge
b. Pivot
c. Condylar
d. Saddle
e. Ball and socket
f. Plane

18.1 a-f Synovial joints

Figure 18.1a-f: Types of synovial joints

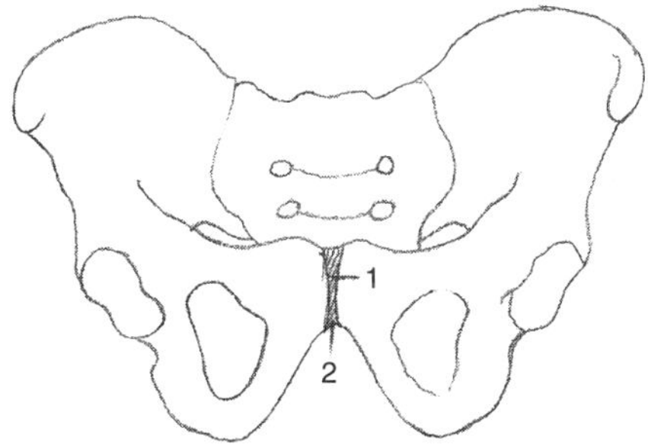

1. Symphysis pubis
2. Fibrocartilage

18.2 Symphysis

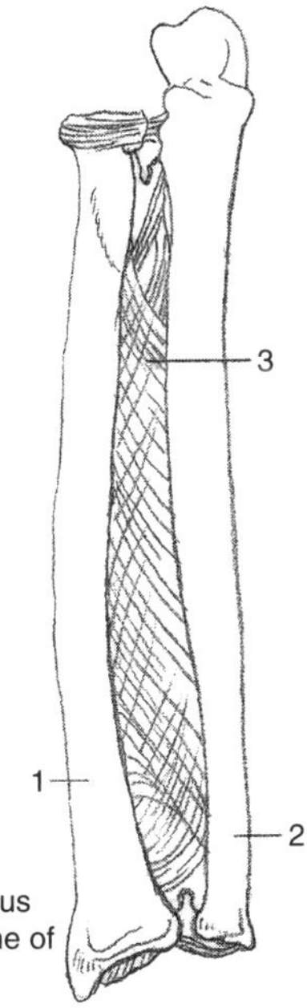

1. Radius
2. Ulna
3. Interosseus membrane of forearm

18.3 Syndesmosis

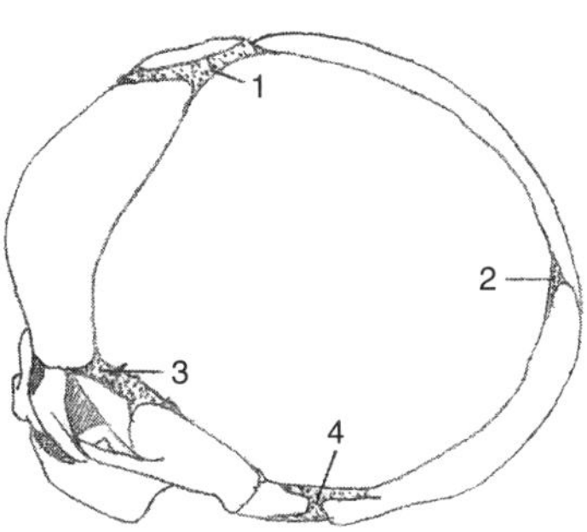

1. Anterior
2. Posterior
3. Anterolateral
4. Posterolateral

18.4 Fontanelles of infant skull

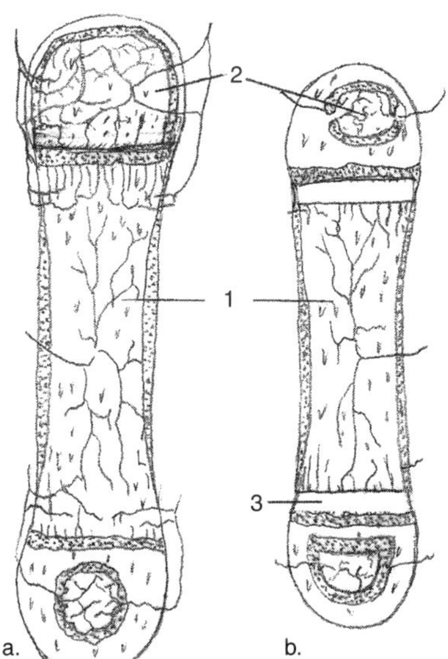

a. b.

1. Diaphysis 2. Epiphysis 3. Cartilage plate

18.5 a-b Synostosis developing from synchondrosis
a. Mature long bone b. Developing bone

Figure 18.2: Symphysis
Figure 18.3: Syndesmosis
Figure 18.4: Fontanelles of the infant skull
Figure 18.5: Synarthrosis

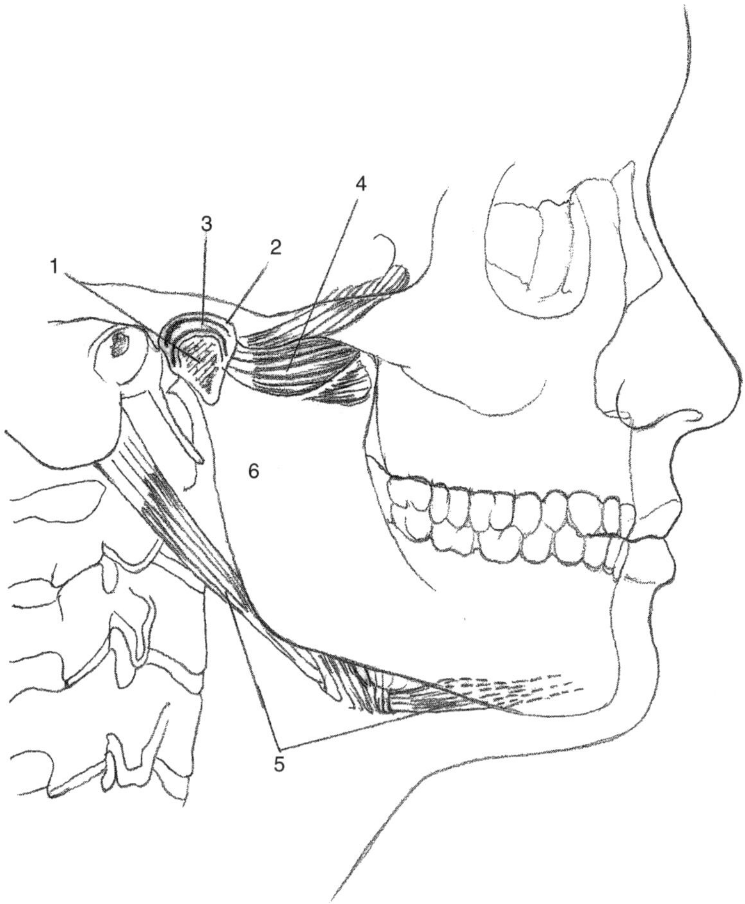

1. Head of mandible
2. Mandibular fossa
3. Joint cavity divided by articular disc
4. Lateral pterygoid muscle
5. Digastric muscle
6. Site of mandibular foramen

18.6 Temporomandibular joint

Figure 18.6: Temporomandibular joint

Chapter 19
FUNCTIONAL ANATOMY OF THE SHOULDER JOINT

The shoulder joint is designed to provide maximum mobility to the upper limb. Its range of motion is approximately 180° in the frontal plane—*i.e.*, the fully adducted limb can be raised high above the head. Perhaps even more impressive is the joint's capacity for circumduction during which the arm traces in circular motion the outline of a cone whose apex is the shoulder (see Fig. 1.4, p. 7).

As you will see, elevation of the upper limb through its complete range requires integrated action at separate and diverse joints. The largest of these, the one most closely identified with the shoulder, is the glenohumeral joint. Others are part of the shoulder girdle joining the clavicle to the sternum and acromion. The contact area between scapula and bony thorax, the scapulothoracic "joint," while not a bony union, enables the scapula to move freely on the chest wall.

The Glenohumeral Joint

The **glenohumeral joint** is a polyaxial, synovial joint of the ball and socket variety (enarthrosis). It may well be the most mobile joint in the body, for the shallow glenoid cavity of the scapula, though deepened in life by the presence of a fibrocartilaginous border, the **glenoidal labrum** (L., *labrum* = lip), makes contact with only about one-third of the near-spherical head of the humerus. It follows, therefore, that bone structure contributes little to the stability of the glenohumeral joint, and a forceful blow to the shoulder is likely to cause joint separation.

The capsule investing the glenoid cavity and upper end of the humerus is solid and substantial. Functional requirements of the joint dictate, however, that the capsule be redundant medially so as not to limit over-the-head abduction of the humerus.

The burden of maintaining joint stability, therefore, falls on muscle; and, in particular, on four short muscles that pass over, in front of, and behind the glenohumeral joint. All originate from the scapula (Fig. 19.1). Supraspinatus, passing to its insertion over the top of the humeral head, is well positioned to raise the humerus from its resting position. Both infraspinatus and teres minor, crossing the joint posteriorly, are able both to extend the arm and to rotate it laterally. Subscapularis, inserting into the lesser tuberosity, is a medial rotator of the humerus. Beyond their essential functions as movers of the arm, these four short muscles act as "guardians of the shoulder joint." Their tendons, blending with the joint capsule to form a musculotendinous cuff, are instrumental in keeping the head of the humerus centered and firmly applied to the glenoid cavity. An aptly descriptive name for the glenohumeral joint capsule is **rotator cuff**—a term consistent with the actions of these muscles and their joint stabilizing functions.

Of course, many other muscles act on the glenohumeral joint; and, in each case, the muscle's relationship to the joint determines its mode of action. Deltoid, for example, whose fibers sweep from a broad origin on the shoulder girdle to a narrow insertion half-way down the humerus, blankets the shoulder joint on three sides (Fig. 19.2); therefore, depending on which part of the muscle is contracting, deltoid can flex, extend, and rotate the arm, both medially and laterally. First and foremost, though, deltoid is a powerful abductor of the humerus. Note, however, that its fibers, being at right angles to the axis of rotation, cannot develop the power (torque) to initiate abduction. Supraspinatus, on the other hand, the short muscle that pulls straight across the head of the humerus, is very important in the early stages of abduction, for it horizontal pull converts the vertical pull of deltoid into rotary motion. Without the assistance of supraspinatus, therefore, deltoid cannot raise the arm from a fully adducted position.

It is possible to conclude from this discussion that a single muscle, such as deltoid, can perform several different functions, yet cannot always get the job done without the assistance of one or more partners. Indeed, movement is produced usually by several muscles acting in concert or in sequence—and shoulder joint dynamics accurately reflect this general principle.

The Scapulothoracic "Joint"

The **scapula** is applied to the back of the thoracic cage and is kept there by the many muscles that crowd onto the bone's borders, angles, and processes. It has anterior (costal) and posterior surfaces that meet along medial and lateral borders. The flat ventral or costal surface of the scapula and the smooth chest wall form a contact area having all the attributes of a multiaxial joint, thus enabling the scapula to move on the thorax in a number of directions.

Motion at the scapulothoracic joint is defined as **protraction** if the scapula moves forwards on the chest wall, and **retraction** if movement is in the opposite direction. The scapula can also be **elevated** and **depressed. Lateral rotation** of the bone, which displaces the inferior angle laterally, causes the face of the glenoid cavity to look up, while **medial rotation** turns it down.

Muscles that produce these movements originate from the axial skeleton. They include: trapezius, serratus anterior, levator scapulae, and romboids. The upper fibers of trapezius, arising from the occipital bone and ligamentum nuchae, descend obliquely to their insertions on the acromion and the lateral one-third of the clavicle (Fig. 19.2). Its lower fibers, originating from the spinous processes of vertebrae T7–12, slant upwards and converge upon the tubercle on the spine of the scapula. Horizontally directed fibers of trapezius attach to the superior lip of the spine between tubercle and acromion. It is easy to imagine, therefore, that the three parts of the muscle, acting independently, can elevate, depress or retract (adduct, brace) the shoulder—actions that are shared, respectively, by levator scapulae, gravity, and the rhomboids. When upper and lower fibers of trapezius act in unison they cause lateral rotation of the scapula, the center of rotation being close to the tubercle on the spine. Without scapular rotation the upper limb cannot be fully abducted, and the two parts of trapezius may be likened to upper and lower components of a force couple; *i.e.*, two equal, but oppositely directed, parallel forces.

Contraction of the serratus anterior muscle, passing between chest wall and the medial border of the scapula's costal surface, causes protraction of the scapula. Moreover, the fibers that reach the inferior angle of the bone become members of the lower force couple in lateral rotation of the scapula.

In medial rotation of the scapula, levator scapulae and pectoralis minor muscles form the upper component of a force couple, and rhomboids its lower component. The horizontal fibers of trapezius act to stabilize the center of rotation whether the scapula rotates medially or laterally.

The Clavicular Joints

The **clavicle**, the other bone of the limb girdle, also may be elevated, depressed, protracted, retracted, and rotated about its long axis.

The **sternoclavicular joint** (Fig. 19.3), the articulation between the medial end of the clavicle and manubrium sterni, is a complex synovial joint; *i.e.*, the joint cavity is divided by a disc of fibrocartilage. Elevation and depression of the clavicle occur mainly between the sternal end of the clavicle and the articular disc, protraction and retraction between the disc and manubrium. The sternoclavicular joint is saddle-shaped, hence structured to allow a limited amount of rotation. As I shall explain, clavicular "spin," equal to some 50°, is an absolute requirement for scapular rotation.

The lateral end of the clavicle and the acromion of the scapula form the **acromioclavicular joint**, a synovial joint of the plane variety. In addition, scapula and clavicle are united by the **coracoclavicular ligament**, a stout cord of densely fibrous connective tissue. This union, classifiable as a syndesmosis, forces scapula and clavicle to move as one, especially during elevation of the clavicle, while giving both bones modest leeway to function independently at other times.

Acromioclavicular joint disorders that freeze the joint interfere with both scapular and clavicular rotation.

Shoulder Joint Dynamics

The muscles acting on the various parts of the shoulder joint complex normally work in perfect balance and harmony.

During the first phase of limb abduction, most of the movement is at the glenohumeral joint. **Supraspinatus** initiates the process by raising the arm some 30°–40° from a fully-adducted position. As the head of the humerus spins in the glenoid cavity, the lateral end of the clavicle and attached acromion are displaced upwards, the sternoclavicular joint acting as fulcrum. The **costoclavicular ligament**, joining first rib and cartilage to the clavicle, limits elevation of the bone to about 50° from the horizontal plane. The scapula, tied to the lateral curvature of clavicle by the coracoclavicular ligament, is raised at the same time, protracted and laterally rotated.

In the second phase of limb abduction, action at the glenohumeral joint continues, but further scapular rotation is resisted by the coracoclavicular ligament. The restraining influence of the ligament is removed, when the clavicle rotates backward about its long axis some 50°, bringing its lateral curvature and the coracoid process closer together (Fig. 19.4). Thereafter, for every 15° of limb elevation, 10° are due to action at the glenohumeral joint and 5° to scapular rotation—with the scapula moving on the clavicle at the acromioclavicular joint. The choreography of motion normally producing effortless upper limb elevation has been dubbed the "scapulohumeral rhythm." Remember, however, only in the presence of fully functional clavicular joints is full range of motion at the shoulder attained; and, when rotation of the clavicle is prevented, the scapular contribution to limb elevation is lost.

Over-the-head abduction of the extremity from a fully adducted position requires lateral rotation not only of the scapula but of the humerus as well. You should verify on yourself that the arm can be raised only to shoulder level when the humerus is medially rotated. This is so, because medial rotation of the bone tightens capsular fibers and prevents the greater tuberosity from clearing the acromial arch (impingement).

As already noted, both supraspinatus and deltoid participate in abduction at the glenohumeral joint, the former initiating the process and the deltoid becoming dominant between 90° and 180° of elevation. For these muscles to be effective, however, their upward pull on the humerus has to be matched by muscles that depress it. **Infraspinatus**, **teres minor**, and **subscapularis** do just that: they form the lower force couple opposing the action of the upper force couple, represented by supraspinatus and deltoid. Without the downward pull of the three short muscles, contraction of supraspinatus and deltoid might easily draw the humeral head out of the glenoid cavity.

An interesting sidelight is the finding that patients incapable of scapular rotation can abduct the limb no more than 90°—though the arm can be lifted passively up to 120° by the physical therapist. Inability to raise the arm the incremental 30° appears to be the result of deltoid's failure to maintain resting fiber length. Simply stated, slack fibers develop little force. And so, when the arm is raised in the absence of scapular rotation, the distance between deltoid's insertion on the humerus and and its origin on the scapula narrows progressively. With the latter's inability to rotate away from the former, deltoid's resting fiber length is not maintained, and the muscle can no longer generate sufficient power to raise the arm past the horizontal.

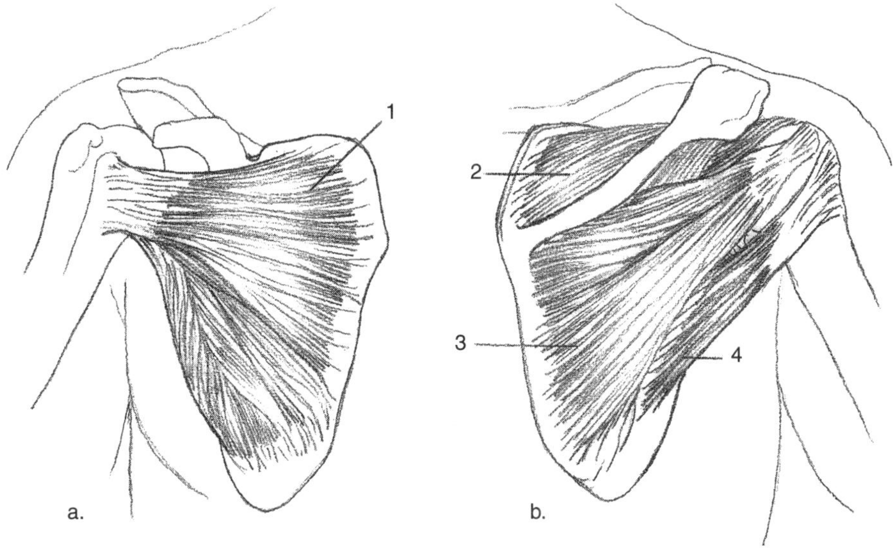

1. Subscapularis
2. Supraspinatus
3. Infraspinatus
4. Teres minor

19.1a-b Scapula and rotator cuff muscles
a. Anterior surface b. Posterior surface

Figure 19.1a-b: Rotator cuff muscles

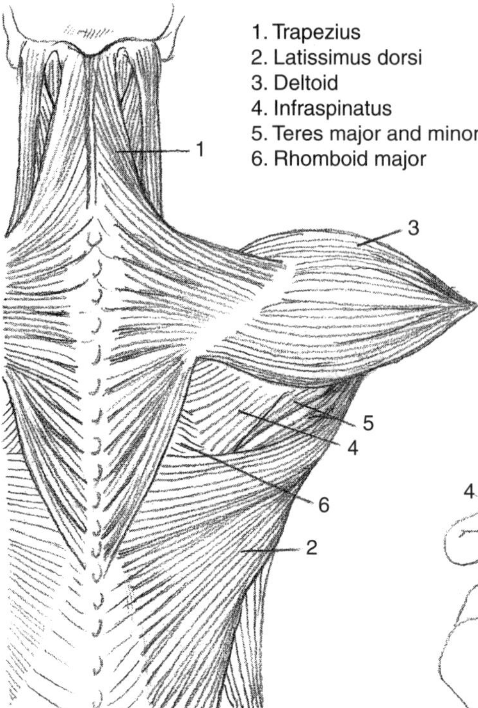

1. Trapezius
2. Latissimus dorsi
3. Deltoid
4. Infraspinatus
5. Teres major and minor
6. Rhomboid major

19.2 Superficial muscles of back and shoulder

1. Sternoclavicular joint cavity and articular disc
2. Costoclavicular ligament
3. Coracoclavicular ligament
4. Acromioclavicular joint

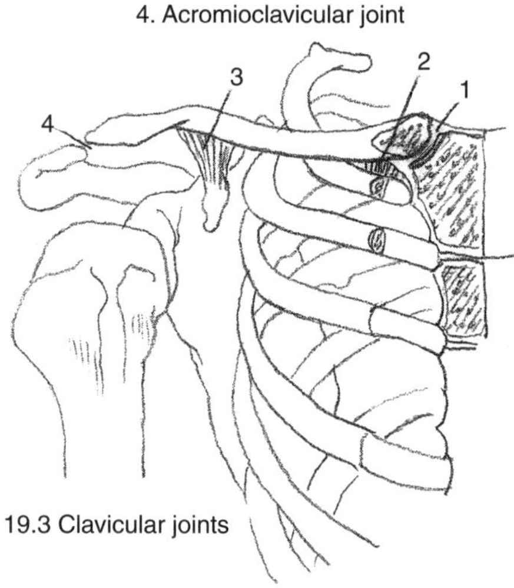

19.3 Clavicular joints

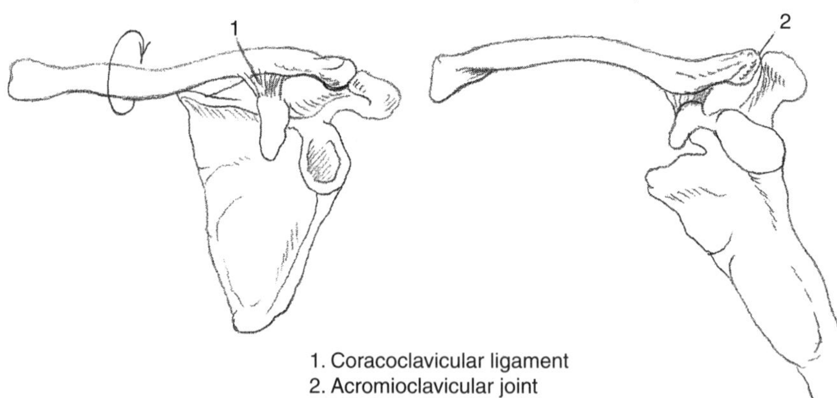

1. Coracoclavicular ligament
2. Acromioclavicular joint

19.4 a-b Coracoclavicular ligament and rotation of clavicle
a. Before b. After rotation of clavicle

Figure 19.2: Superficial muscles of the back and shoulder
Figure 19.3: Clavicular joints
Figure 19.4 a-b: Effect of clavicular rotation on the scapula
a-Before rotation
b-After rotation

Chapter 20
INNERVATION OF UPPER LIMBS

This chapter begins with a short review of human limb development, and of the evolutionary process that causes flexor and extensor surfaces of upper and lower limbs to become transposed—thus, to face in opposite directions. It then discusses the composition and functional organization of the brachial plexus, refers to brachial plexus injuries, and comments on the cutaneous innervation of the upper extremities. A thorough description of the lumbosacral plexus and its branches appears in chapter 32, "Innervation of Lower Limbs."

Limb Development

The limb primordia, known as **limb buds**, can be recognized early in the fifth week of human life as paired lateral outgrowths from the body wall in the future neck and abdominopelvic regions. The upper limb buds appear somewhat sooner than the lower pair; but, by the end of the week, proximal and distal limb segments are clearly evident in both (Fig. 20.1).

At the tip of each limb bud is the **apical ectodermal ridge** (AER). The prevailing view is that AER induces mesenchymal cell in the limb bud's core to differentiate into chondrocytes, from which cartilage models of the future appendicular skeleton will form. Other undifferentiated mesenchymal cells become myoblasts which, *in situ*, develop into skeletal muscle.

Upper and lower limb buds, respectively, lie opposite several myotomes in the neck and abdominopelvic regions. Each myotome is invaded by outgrowing spinal nerves early in development; and, when myoblasts from cervical and lumbosacral myotomes migrate into the substance of the limb bud, they drag nerve fibers of several spinal cord levels along. The upper pair of limb buds receives fibers from segments C5 through T1, and the lower pair from segments L2 through S3.

Early on, the future limbs project at right angles from the embryo's trunk -like oars on a row-boat (Fig. 20.2). Hence, they have anterior and posterior surfaces that meet along superior and inferior borders. Note, that the terms *superior* and *inferior* are descriptive of the limb's borders only so long as they remain perpendicular to the body's long axis. Terminology that defines the borders with respect to the central axis of the limb, *i.e.*, **pre-axial** for superior and **postaxial** for inferior, is more meaningful once the limbs undergo rotation and folding. Thus, thumb and great toe both are located at the pre-axial (superior) borders of the limb buds, the little finger and toe at the postaxial (inferior) borders.

Superficial veins serve as guides to borders in the fully developed limb. Cephalic and basilic veins, respectively, map out pre- and postaxial borders of the upper extremity. In the lower limbs, the great (long) saphenous vein lies on the pre-axial border, and the small saphenous vein runs along the postaxial border—though the arrangement here is obscured owing to rotation and extension of the lower extremities in later development.

Pre- and postaxial borders are connected internally by a fetal intermuscular septum; it divides the limb bud into anterior and posterior compartments (Fig. 20.3). Flexor muscles of the limb will develop in the anterior compartment, and extensor muscles in the posterior compartment. Innervation is supplied by a limb plexus whose trunks separate into anterior and posterior divisions. The **anterior divisions** supply the flexor (anterior) compartment of the limb, and the **posterior divisions** its extensor (posterior) compartment. It is noteworthy that the flexor compartment has a richer nerve supply; hence, flexor muscles are faster acting and more versatile than extensors, and flexor skin is more sensitive than extensor skin, especially distally in the limb. Also, the lowest (most postaxial) segmental nerve in the plexuses—*i.e.*, T1 and S3 in brachial and lumbosacral plexus, respectively, distributes only to the flexor compartment.

During the second month of human development, both upper and lower extremities fold towards the front; flexion at the elbow, as well as of the hip and knee, causes them to assume typical fetal positions. Concurrently, both rotate about their long axes; but, and mark this well, the upper limbs undergo lateral rotation and the lower limbs, rotate medially. Therefore, when the limbs unfold, the pre-axial (thumb) borders of the upper pair come to lie on the outer side of the limb and the pre-axial (great-toe) borders of the lower pair on the inner side of the limb. The postaxial borders of upper and lower limbs, likewise, become transposed to opposite sides. Equally important, counterdirectional rotation reverses limb surfaces and compartments. The results are especially significant in the lower extremities, whose flexor compartments—innervated by anterior divisions of the limb plexus—are transported to the back of the limbs, and whose extensor compartments—supplied by posterior divisions of the limb plexus—come to occupy the front of the limbs.

The Brachial Plexus

The brachial plexus supplies sensory, motor, and autonomic (visceral motor) nerve fibers to the upper limb. Five distinct, programmatic pieces go into the making of the brachial plexus (Fig. 20.4), each step in the program carrying a very specific and proprietary designation: (1) **Roots** are the anterior rami of spinal nerves C5–T1. (2) **Trunks** are formed by the unions of roots C5–6 (upper trunk) and C8–T1 (lower trunk), with root C7 continuing as middle trunk. (3) **Divisions,** anterior and posterior, result as each trunks splits. (4) **Cords** are named according to their relationship to the second part of the axillary artery: Medial and lateral cords are derived from anterior divisions and the posterior cord is formed by posterior divisions. (5) **Terminal branches** are the final pathways to arm, forearm, and hand.

In short, the five roots of the plexus yield five named terminal branches; but, whereas the roots are segment-specific, the terminal branches carry fibers from several spinal cord segments.

Nerves arising from the brachial plexus proximal to the terminal branches are designated **collateral branches**. Some come off roots and trunks (supraclavicular collaterals) and others split off the cords (infraclavicular collaterals). Most, like the terminal branches, are multisegmental nerves.

Functional Organization of the Brachial Plexus

In general, the higher placed (pre-axial) spinal nerves in the brachial plexus supply muscles situated proximally in the limb, and the lower placed (postaxial) spinal nerves distribute to more distally located muscles. Thus, anterior rami of C5–6 supply muscles of the shoulder girdle (*e.g.*, rhomboids, levator scapulae) and arm (*e.g.*, deltoid, biceps). Anterior rami of C7–8 supply forearm muscles, and anterior rami of C8–T1 distribute to the muscles of the hand. Be forewarned, however, the cutaneous distribution of brachial plexus nerves is patterned differently.

Remember that all muscles developed in the primitive posterior compartment of the upper limb are supplied by posterior divisions of the brachial plexus. The posterior divisions join to form the posterior cord, whose terminal branches are **radial** and **axillary nerves**, and whose collaterals are **thoracodorsal** and **subscapular nerves** (first letters spell RATS). The radial nerve supplies *all* posterior compartment muscles of arm and forearm, *i.e.*, the extensors of the elbow, wrist, and digits. The other terminal branch, the axillary nerve, innervates *only two muscles*— deltoid and teres minor. Thoracodorsal and subscapular nerves distribute to the muscles of the posterior axillary fold, latissimus dorsi and teres major, and to subscapularis—one of three muscles that clothe the part of the scapula developing posterior to the midaxillary line. The other two muscles, supra- and infraspinatus, are innervated by the **suprascapular nerve**.

The anterior compartment of the upper limb receives innervation from anterior divisions of the brachial plexus joining to form lateral and medial cords. Their terminal branches, **musculocutaneous, median**, and **ulnar nerves**, supply the flexors of the elbow, wrist, and digits. Collateral branches, *e.g.*, **lateral** and **medial pectoral nerves**, innervate other muscles that developed morphologically anterior to the skeletal axis of the primitive upper limb and in front of the midaxillary line; namely, the muscles of the anterior axillary fold and those arising from the coracoid process of the scapula. The distribution of the three terminal branches follows a simple plan: muscles in the flexor compartment of the arm are supplied by the musculocutaneous nerve, those in the flexor compartment of the forearm are innervated by the median nerve, and all muscles of the hand by the ulnar nerve. That is the general principle, and here are the exceptions: (1) *1+1/2 muscles in the forearm are innervated not by the median but by the ulnar nerve*; they are the flexor carpi ulnaris, plus the two bellies of the flexor digitorum profundus on the medial (ulnar) side. And (2), using the same magic number, *1+1/2 muscles in the hand are innervated, not by the ulnar nerve, but by the median.* I am taking the liberty here of considering the muscles of the thenar eminence—flexor and abductor pollicis brevis, and opponens pollicis—to be a unit of one, and the 1/2 muscle to stand for the two lumbricals on the lateral (radial) side of the hand.

Brachial Plexus Injuries

Obstetrical paralysis, affecting shoulder and arm muscles, refers to sequelae resulting from damage to the upper trunk of the plexus (C5–6) when the head of a presenting infant is forcibly rotated before the shoulder clears the birth canal. The lower trunk (C8, T1) is likely to be injured, causing paralysis of the small muscles of the hand, when excessive traction is applied to the outstretched limb—as happens when a child is swung in a circle by its arms.

As the neurovascular bundle for the upper limb leaves the root of the neck, it may become entrapped between scalenus anterior and medius muscles, encounter a cervical rib, or pass through a costoclavicular interval constricted by an abnormally high first rib. Commonly observed results of compression syndromes include vascular insufficiency, including faint pulse, and impaired function of the muscles of the hand. Likewise, compression of the median nerve in the carpal tunnel produces weakness of thenar muscles and sensory loss in the territory of the nerve's distribution.

Damage to other parts of the brachial plexus or its branches causes predictable deficits in upper limb function, including a characteristic inability to maintain the hand in a "neutral" position.

Cutaneous Innervation of the Upper Limb

A **dermatome**, as you know, is defined as an area or patch of skin supplied by a single spinal nerve, corresponding to a single spinal cord segment. The dermatomes encircling the trunk form narrow, horizontal bands (see Fig. 8.3, p. 54), representing, in numerical order, spinal nerves T2 though L1; *i.e.*, the segmental outflow between limb plexuses. Along the limb also, dermatomes are arranged in orderly sequence—perhaps difficult to recognize at first, but plainly obvious once you remember that the developing limb bud carried the body wall skin opposite spinal cord segments C5–T1 with it. The higher segments whose nerves enter into the formation of the brachial plexus distribute to skin along the pre-axial border of the limb, from proximal to distal, in numerical order (Fig. 20.5). The lower segmental nerves of the plexus innervate the skin along the postaxial border of the limb also in ascending numerical order, but from distal to proximal. Accordingly, dermatomes C5 and T2 lie next to each other at the root of the limb, and the centrally placed nerve in the plexus, C7, distributes to the skin most distal in the limb.

Considerable overlap in cutaneous distribution exists where numerically ordered dermatomes meet. There is no overlap, however, of dermatomes separated by axial lines as shown in Figure 20.5. Therefore, to pin-point nerve root damage, you must test across axial lines.

The distribution of a named cutaneous nerve of the upper limb is clearly consistent with the segmental outflow of fibers carried by that nerve. Thus, the lateral cutaneous nerves, whose fibers come from spinal cord segments C5–6, innervate the skin over the limb's pre-axial border; whereas the medial cutaneous nerves of arm and forearm, both containing fibers from spinal cord segments C8–T1, supply the skin over the postaxial border of the limb. Conversely, of course, knowing the segmental outflow of a given cutaneous nerve, lets you predict its distribution.

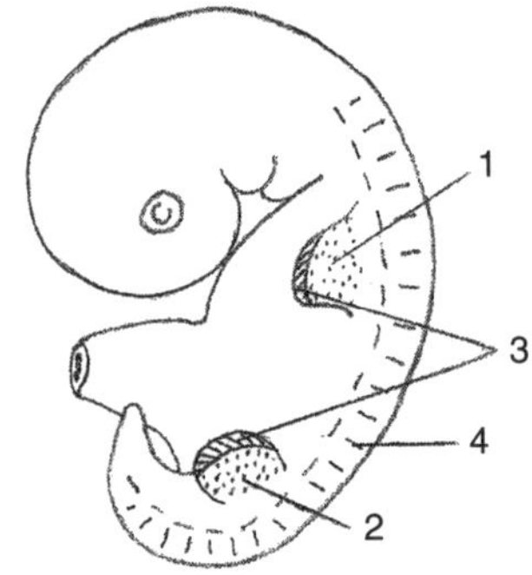

1. Anterior limb bud
2. Posterior limb bud
3. Apical ectodermal ridge (AER)
4. Somites

20.1 Limb bud stage

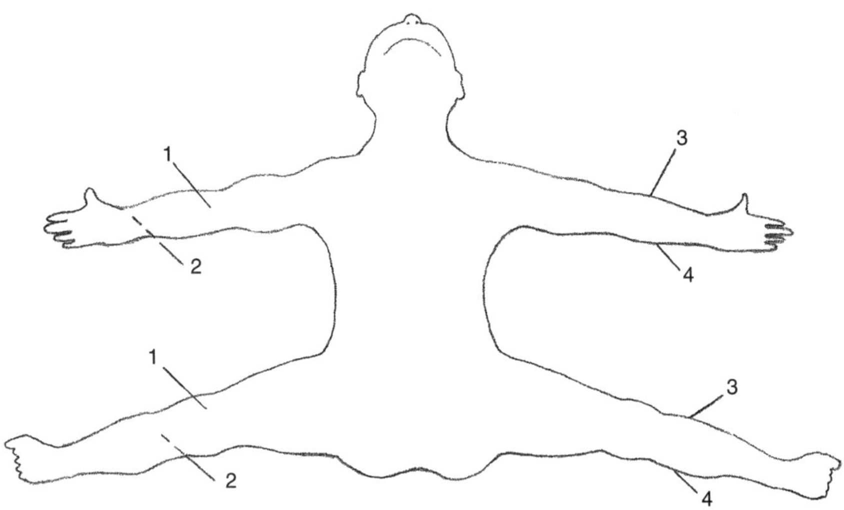

1. Anterior (ventral) surface
2. Posterior (dorsal) surface
3. Superior (pre-axial) border
4. Inferior (postaxial) border

20.2 Surfaces and borders of fetal limb

Figure 20.1: Limb bud stage
Figure 20.2: Surfaces and borders of the fetal limb

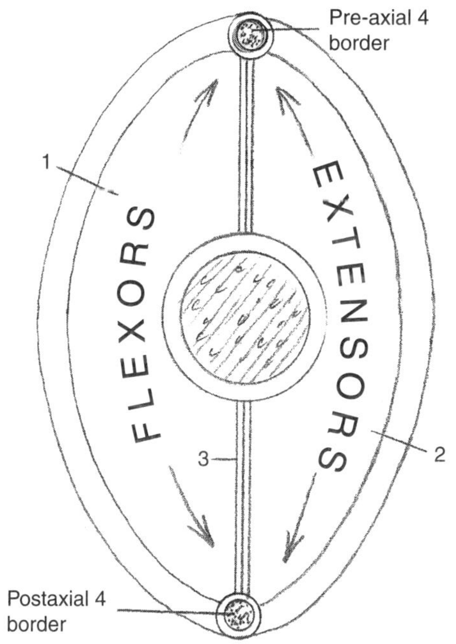

1. Anterior (flexor) compartment
2. Posterior (extensor) compartment
3. Intermuscular septum
4. Superficial vein at pre- and postaxial borders

20.3 Fetal limb with compartments (cross-section)

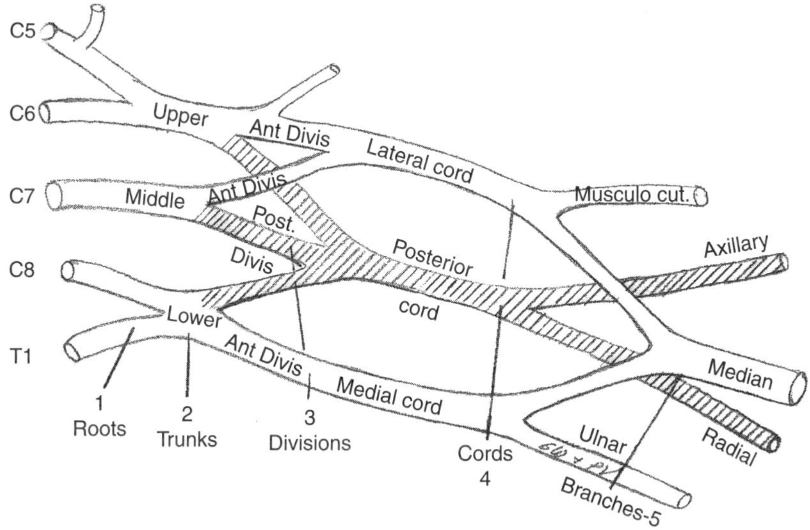

1. Roots (C5-8, T1)
2. Trunks (Upper, Middle, Lower)
3. Divisions (Anterior and Posterior {hatched})
4. Cords (Lateral, Medial, and Posterior {hatched})
5. Terminal branches (Musculocutaneous, Median, Ulnar, and Axillary, Radial {hatched})

20.4 Anatomy of brachial plexus

Figure 20.3: Cross-section of fetal limb with compartments
Figure 20.4: Anatomy of the brachial plexus

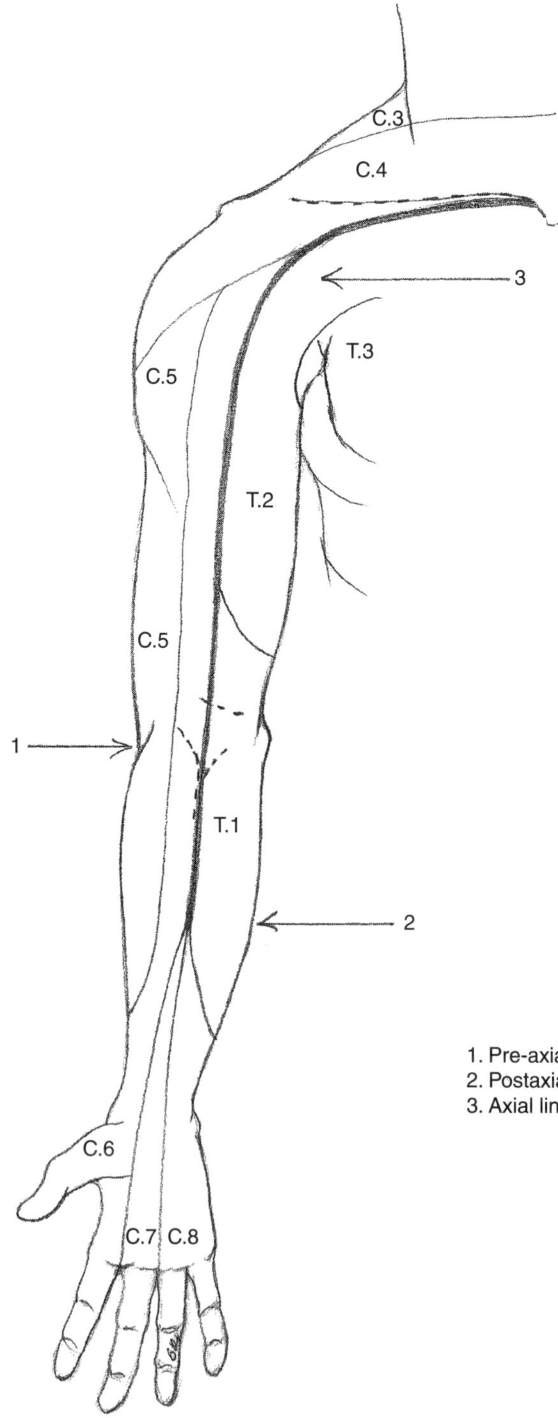

1. Pre-axial border
2. Postaxial border
3. Axial line

20.5 Cutaneous innervation of upper limb

Figure 20.5: Cutaneous innervation of the upper limb

Chapter 21
THE HAND

The hand is an organ of great functional diversity and its anatomy, therefore, is complex. It is crammed full of small muscles and packed with tendons descending from the forearm. Two major arterial anastomoses form arcades or arches across the width of the hand, and many slender branches of two large nerves supply its muscles and skin. The basic anatomy of the hand may be made clearer by drawing a cross-section of the palm, but first a few words about skin and palmar aponeurosis.

Skin

The skin of the hand is highly sensitive—especially that at the tip of the fingers, described by some as the "eyes of the hand." Tactile acuity, *skin sense*, clearly is the essential ingredient of the hand's exteroceptive function.

On the palmar (ventral or volar) aspect of the hand, the skin is firmly bound to the underlying palmar aponeurosis—bonding that minimizes slippage of skin on subcutaneous tissue. On the dorsum, however, the skin can be picked up and moved easily. The fascia is thin and extensor tendons can be seen playing beneath it. Observe that veins draining blood from the hand pass to the dorsal rather than to the palmar surface, where they would be subject to compression when the hand is clenched.

The palmar aponeurosis takes the place of superficial fascia. It is triangular in shape, its apex pointing towards the wrist. The base of the palmar aponeurosis gives off four fibrous slips, which help form **fibrous digital sheaths**—interrupted tunnels of dense connective tissue through which long digital flexor tendons run to their insertions (Fig. 21.1). When present, **palmaris longus**—a muscle undergoing phylogenetic degeneration—is continuous with the apex of the palmar aponeurosis. The muscle, acting through the palmar aponeurosis and fibrous digital sheaths, can flex the fingers. So will pathologic contraction of the palmar aponeurosis: Dupuytren's contracture refers to a clinical condition characterized by permanent flexion of the fingers, especially digits IV and V.

Palmar Compartments

A cross-section of the palm reveals thenar, central, and hypothenar compartments—separated by deep fascia (Fig. 21.2).

The **thenar compartment** holds the three muscles of the thenar eminence at the root of the thumb or **pollex** (L., *Polleo* = I am strong)—flexor and abductor pollicis brevis and opponens pollicis. All are innervated by the recurrent branch of the median nerve, a branch covered only by skin and deep fascia, hence easily traumatized. Compartment skin is innervated by spinal nerve C6.

The **hypothenar compartment** contains the three muscles of the hypothenar eminence at the base of the little finger—flexor digiti minimi brevis and abductor and opponens digiti minimi. These muscles are supplied by the ulnar nerve, as are all others in the hand except for thenar muscles and lumbricals I and II. Cutaneous innervation is via spinal nerve C8.

The **central compartment** has anterior and posterior walls. The anterior wall is formed by the palmar aponeurosis which, along the sides, blends with the deep fascia over thenar and hypothenar muscles. The posterior wall is composed of fascia, interosseus muscles, and metacarpal bones. The intermediate (oblique) palmar septum, passing between metarpal bone III and palmar aponeurosis, divides the central compartment into **medial** (ulnar) and **lateral** (radial) **midpalmar spaces**. The transverse head of the adductor pollicis muscle lies in the floor of the lateral midpalmar space, superficial to interosseus muscles and metacarpal bones.

The laminar fascia forming the posterior wall of the central compartment extends proximally into the anterior compartment of the forearm, where it becomes the anterior wall of Parona's space, bounded posteriorly by the pronator quadratus muscle and fascia. The space is clinically important, as you can guess, because infections originating in the hand may spread by this route into the forearm.

The central compartment is occupied by units of long flexor tendons in **synovial digital sheaths**. Three flexor units lie within the medial midpalmar space; and one, the unit to the index finger, is located in the lateral midpalmar space. Each unit is related on its radial side to a digital artery, a digital nerve, and a **lumbrical muscle** inside a fascial canal (L., *lumbricus* = a worm). In the palm, the order from superficial to deep is: artery, nerve, muscle. Spinal nerve C7 innervates the skin of the central compartment.

Synovial Sheaths

The tendons of the long digital flexors, *i.e.*, flexor digitorum superficialis and flexor digitorum profundus—but other flexor, extensor, and abductor tendons as well—are wrapped in synovial membrane. **Synovial digital sheaths** (Fig. 21.3), inside fibrous digital sheaths, surround superficial and deep flexor tendons of each finger. In the wrist and the palm, the long tendons are enclosed in **carpal tendinous sheaths,** of which there are two (or three, if you include the sheath of flexor carpi radialis): the **common flexor sheath**, also known as the **ulnar bursa**, accomodates the four tendons each of flexors digitorum superficialis and profundus; the **tendinous sheath of pollicis longus**, or **radial bursa**, invests the long tendon of the thumb.

Frequently, ulnar and radial bursae communicate. Moreover, the synovial digital sheath of the thumb virtually always joins the radial bursa, and the synovial digital sheath of the little finger commonly links up with the ulnar bursa. In such cases, an infection caused by a penetrating wound into the synoval digital sheath of the thumb or little finger may become a "horse-shoe" abcess.

Carpal tendinous sheaths project proximally into the **carpal tunnel**, the osseofascial passage for the long flexor tendons and the median nerve. The proximal limits of ulnar and radial bursae are indicated on the wrist's anterior aspect by the middle skin crease (among the three that are usually present); it also is a fairly reliable surface marker of the wrist (radiocarpal) joint space. The distal limits of carpal tendinous sheaths are roughly opposite and deep to the transverse skin creases in the palm of the hand.

The **flexor retinaculum** (transverse carpal ligament) is a densely fibrous band attached to the outer four carpal bones. It forms the roof of the carpal tunnel, and keeps structures descending into the palm locked down. The ulnar nerve, artery, and the tendon of palmaris longus pass superficial to the flexor retinaculum, and most of the thenar and hypothenar muscles take origin, at least partially, from its anterior surface. The proximal border of the flexor retinaculum and, therefore, of the carpal tunnel lies deep to the prominent distal skin crease at the wrist.

Lumbrical Muscles

Lumbricals I and II arise from flexor digitorum profundus tendons I and II, going to index and middle fingers, respectively. Lumbricals III and IV originate by two heads from neighboring profundus tendons: the former from tendons II and III, the latter from tendons III and IV. Their innervation corresponds to that of the parent deep flexor muscle.

The four lumbricals pass to the radial side of the four fingers, cross a metacarpophalangeal joint from anterior to posterior, and insert distally into the **dorsal digital** (extensor) **expansion** of the same-numbered extensor digitorum tendon (Fig. 21.4). All, therefore, are able to flex the metacarpophalangeal joints.

Interosseus Muscles

Interosseus muscles (Fig. 21.5) abduct and adduct the digits—the middle finger being defined as the central axis. All arise from the sides of metacarpal bones, all cross metacarpophalangeal joints, and all insert into the base of proximal phalanges *and* into the dorsal digital expansion in common with a lumbrical muscle but proximal to it.

The four palmar interosseus muscles are adductors (**P-ad**) and the four dorsal interossei are abductors (**D-ab**) of the digits. Each **palmar interosseus muscle** originates from the metacarpal bone of the digit which it moves: interossei I and II from the medial aspect of the bone, and interossei III and IV from the lateral aspect. None attaches to the middle finger. The first palmar interosseus muscle is merely an accessory to the powerful adductor pollicis.

The **dorsal interosseus muscles** lie deep to the palmar interossei and differ from them also in being bipennate; *i.e.*, they take origin from the sides of adjoining metacarpal bones. Dorsal interosseus I, for example, the abductor of the index finger, arises from the sides of metacarpals I and II. Similarly, interosseus muscle IV, which acts to abduct the ring finger, originates from metacarpals IV and V. Notice that neither the thumb nor the little finger requires dorsal interossei; their abductors are part of the thenar and hypothenar musculature. Finally, be aware that the radial artery enters the palm by passing between the two heads both of dorsal interosseus I and of adductor pollicis.

The Interosseus-Lumbrical Complex

Interosseus muscles are capable not only of abducting and adducting the digits, but also of flexing and extending them by acting in concert with lumbricals. The muscles of the interosseus-lumbrical complex contribute to, and insert into, the dorsal digital expansion through which they operate when others cannot perform their functions to best advantage. Keep in mind that the long digital flexor and extensor muscles originate from humeral epicondyles; having to travel over distance and across several joints, *e.g.*, elbow and wrist, diminishes their efficiency as flexors and extensors of the digits.

Next, consider this: the long digital flexors do not insert into the proximal phalanges and, therefore, cannot act directly on the metacarpophalangeal joints. As for the extensor digitorum, the deep portion of its tendon inserts into the base of the proximal phalanx, while the superficial part attaches to the base of the middle phalanx (Fig. 21.4). In crossing the metacarpophalangeal joint (the "knuckle"), each tendon spreads out to create the dorsal digital expansion, from which two lateral tendinous slips continue to the base of the distal phalanx.

The muscles of the interosseus-lumbrical complex aid and abet the functions of the long flexors and extensors. The interosseus muscles, in particular, being attached to the dorsal digital expansion distal to the metacarpophalangeal joint, are able to flex that joint. And the pull of lumbricals on the lateral slips of the expansion extends both (proximal and distal) interphalangeal joints when the digit's proximal phalanx is fully extended by the deep portion of the extensor digitorum tendon. In sum, the muscles of the interosseus-lumbrical complex flex the metacarpophalangeal and extend the interphalangeal joints. Their actions are prime factors in manual dexterity, because they function where long flexors and exensors cannot do so effectively.

One of the fundamental principles of muscle physiology bears repeating here: neither overstretched nor slack muscles develop maximal tension; they contract most forcefully when their fibers are at resting length. Actually, fibers at resting length inside the body typically are under slight stretch—to prove this statement correct, detach a muscle from bone and watch it shorten. The relationship between fiber length and contractile force also explains why wrist extension, which tenses the long flexor muscles, ensures a tight grip on a tennis racket or baseball bat. Conversely, wrist flexion leaves the tendons lax and the hand ill-suited to grip firmly.

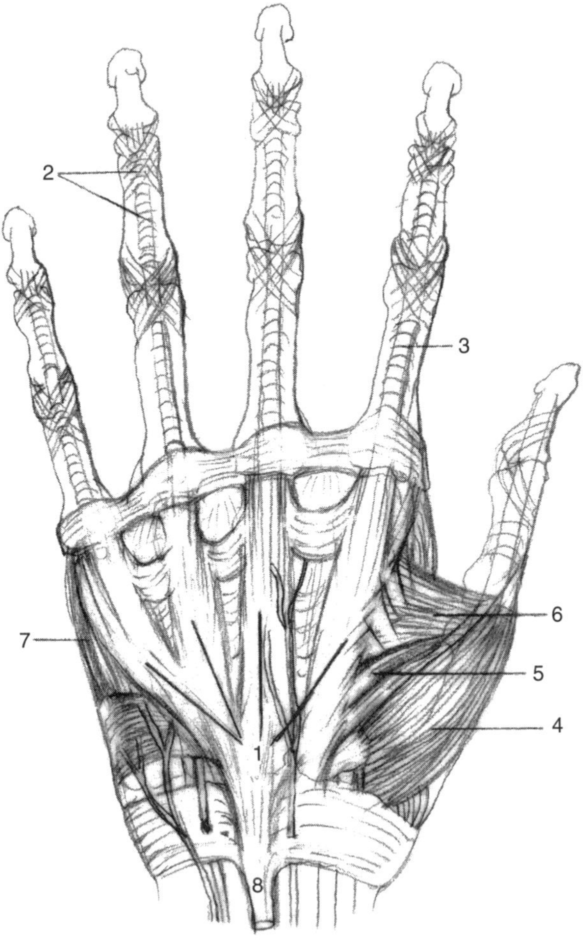

1. Palmar aponeurosis
2. Fibrous digital sheath
3. Long digital flexor tendon
4. Abductor pollicis brevis m.
5. Flexor pollicis brevis m.
6. Adductor pollicis m.
7. Abductor digiti minimi m.
8. Tendon of palmaris longus m.

21.1 Palmar aponeurosis

Figure 21.1: Palmar aponeurosis

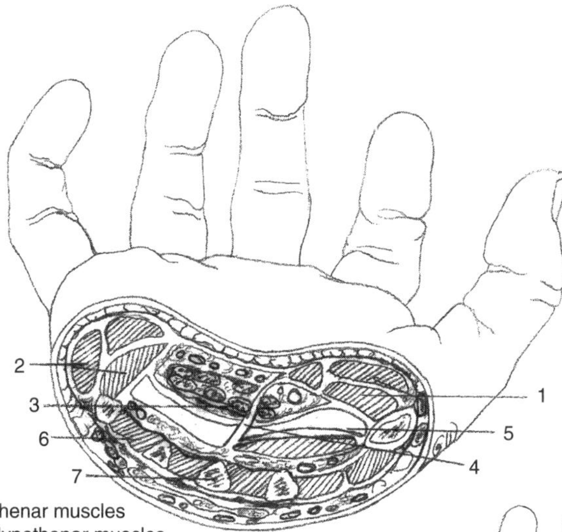

1. Thenar muscles
2. Hypothenar muscles
3. Central compartment with leader line passing through (expanded) medial midpalmar space
4. Intermediate palmar septum
5. Lateral midpalmar space (expanded)
6. Interosseus muscle
7. Metacarpal bone

21.2 Palm of hand (cross-section)

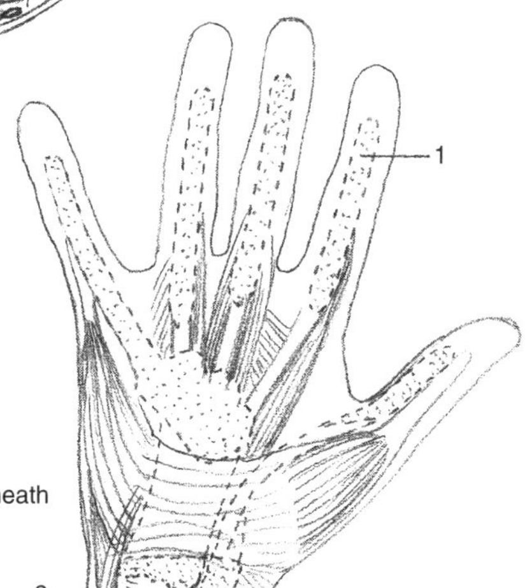

1. Synovial digital sheath
2. Ulnar bursa
3. Radial bursa

21.3 Synovial sheaths

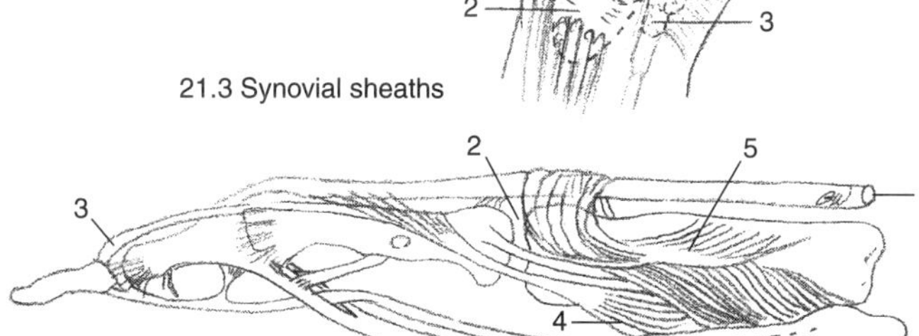

1. Extensor digitorum tendon
2. Metacarpophalangeal joint crossed by tendons of interosseus and lumbrical mm.
3. Tendinous slip of expansion
4. Lumbrical muscle
5. Interosseus muscle

21.4 Dorsal digital expansion (lateral view)

Figure 21.2: Palm of hand (cross-section)
Figure 21.3: Synovial sheaths
Figure 21.4: Dorsal digital expansion (lateral view)

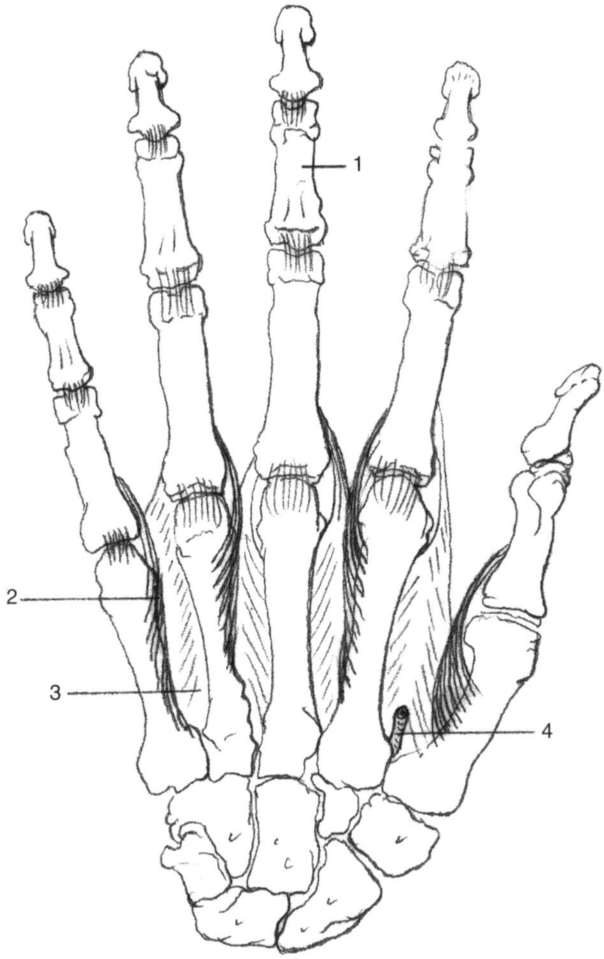

1. Axial digit
2. Palmar interosseus muscle (shaded)
3. Dorsal interosseus muscle
4. Radial artery

21.5 Functional anatomy of interosseus muscles

Figure 21.5: Functional anatomy of interosseus muscles

Chapter 22
THE ANTERIOR ABDOMINAL WALL

The subcostal body wall is largely muscular in its overall construction. The muscles not only function in their usual capacity as movers of the skeleton, *e.g.*, they flex and rotate the trunk, but also provide support and protection for abdominal contents. In addition, a muscular body wall is able to accomodate to variations in intra-abdominal pressure and to comply with changes in size, shape, or position of internal organs, *e.g.*, an enlarging uterus of pregnancy.

We'll first explore the anatomy of the body wall with the aid of a schematic horizontal section through the trunk, then look closely at the inguinal region, a clinically important site of inguinal hernias.

Myofascial Components

The anterolateral wall of the abdominal cavity is formed by paired flat muscles, three layers deep, which in their triple-decker arrangement, terminology, and fiber orientation resemble the muscles of the thoracic cage (Fig. 22.1). They are, from superficial to deep: the **external oblique**, the **internal oblique**, and **transversus abdominis**. **Transversalis fascia** lines the internal aspect of transversus abdominis; it corresponds to the endothoracic fascia and is continuous with it through openings in the diaphragm. Transversalis fascia rests upon parietal peritoneum as endothoracic fascia does on parietal pleura, and the peritoneal cavity is to the abdomen what pleural cavities are to the thorax.

A fourth pair of muscles, **rectus abdominis**, passes vertically in the anterior abdominal wall between costal margin and pubis. Note the presence of three horizontal tendinous inscriptions; they probably represent fusion lines of segmental myomeres that consolidated in development (Fig. 22.2).

The paired flat muscles arise from bone, cartilage, and ligaments framing the subcostal region; and, having nothing solid to attach to in front, their aponeuroses join across the midline to form the double-barreled **rectus sheath** (Fig. 22.1). The **linea alba** is the seam where the aponeuroses of partners from opposite sides became stitched together. Rectus abdominis, contained in an aponeurotic sleeve, can now perform its functions without interference from muscles whose directional pull is perpendicular to its own.

Neurovascular Components

The muscles of the anterolateral body wall are innervated segmentally by anterior rami of the lower six thoracic nerves (intercostal nerves 7–11 plus subcostal) and nerve L1.

The blood supply is by way of **superior** and **inferior epigastric arteries**. The former, a descending terminal vessel of the internal thoracic artery, anastomoses with the latter, an ascending branch of the external iliac artery, on the posterior surface of rectus abdominis inside the rectus sheath. Their connection establishes an end-to-end anastomosis in the anterior abdominal wall, a potential arterial shunt between upper and lower extremities. The lower intercostal and the lumbar arteries complete the vascularization of the anterolateral body wall.

The neurovascular plane is situated between internal oblique and transversus abdominis muscles.

The Inguinal Region

The external oblique is the outermost of the three flat muscles. It originates laterally from the lower eight ribs by fleshy slips which interdigitate with similar slips from latissimus dorsi and serratus anterior muscles. Below the level of the **anterior superior iliac spine** (ASIS), the fleshy part of the muscle is replaced by an aponeurosis whose lower, taut edge is free except for attachments to the ASIS laterally and the pubic bone medially (Fig. 22.2). This lower, free, aponeurotic border of the external oblique is the **inguinal ligament**—not so much a ligament in the conventional sense as a margin of aponeurosis. Its divided attachment to the tubercle and crest of the pubic bone leaves a triangular gap—the **superficial inguinal ring**. Emerging from the ring in the male is the **spermatic cord**; in the female, the **round ligament of the uterus**; and, in both sexes, the **ilio-inguinal** nerve—a major branch of L1.

The superficial inguinal ring is the external aperture of the **inguinal canal**, whose internal aperture, the **deep inguinal ring**, is buried in extraperitoneal connective tissue, including transversalis fascia. In the fully grown male, the surface projection of the inguinal canal may be indicated by a line drawn parallel to, but somewhat above, the inguinal ligament. It starts roughly 13 cm (1/2 in.) lateral to the midinguinal point (located equidistant between ASIS and pubic symphysis) and ends at the superficial ring. The canal, therefore, projects anteriorly, inferiorly, and medially.

The inguinal canal is not, I repeat, *is not* a tube having rigid walls and fixed dimensions. It *is* an ill-defined and obliquely-placed interval between and among muscles, aponeuroses, and fasciae of the lower abdominal wall, and compressed by them into a narrow, slit-like passage for the spermatic cord and round ligament of the uterus.

Developmental Anatomy of Inguinal Region

The inguinal region, or **groin**, is a weak spot in the lower abdominal wall, and especially so in the male. To make sense of the inguinal region, you must go back to its development in the embryo and fetus; for, the anatomy of the inguinal region is shaped by testicular (and ovarian) descent during the second one-half of the fetal period.

The human gonads develop retroperitoneally, *i.e.*, *behind* the primitive body cavity, or celom, high up on the posterior body wall of the embryo (Fig. 22.3a). They are held in place, for the time being, by ligamentous condensations of extraperitoneal connective tissue. One ligament, anchoring the upper pole of the organ, carries gonadal vessels and, in the male, the future sperm-delivery system—the **ductus deferens**. The other, known as **gubernaculum**, connects the lower pole of the gonad with swellings below destined to become scrotum or its homologue in the female, the labia majora. As the fetal period begins, the testis—and let us concentrate on the testis for now—starts migrating towards its ultimate destination, the scrotum. Contributing to testicular descent, undoubtedly, are growth and elongation of the trunk and relative shortening and widening of the gubernaculum testis. By the end of the third month the retroperitoneal testis lies at the junction of posterior and anterior adominal walls. Meanwhile, the scrotum has been developing, and it is invaded by a diverticulum of peritoneum lining the celom. This is the **processus vaginalis peritonei** (Fig. 22.3b). Descending behind it in extraperitoneal connective tissue, and using the gubernaculum as a guide rope, is the testis. As the processus vaginalis passes through the fetal multi-layered abdominal wall with the testis in pursuit, both invaginate first the transversalis fascia, then the developing muscle of the inguinal region. Along the way, processus vaginalis, the testis, and all its dependent structures become concentrically wrapped in tubular prolongations of each layer. The membranes so formed are, from inside out, (1) the **internal spermatic fascia**, a continuation of transversalis fascia, (2) the **cremaster muscle**, derived from internal oblique fibers and (3) the **external spermatic fascia**, representing the external oblique muscle. (Transversus abdominis falls short of the inguinal region). These three coats are the coverings of the spermatic cord, containing the processus vaginalis peritonei, ductus deferens, and testicular vessels and nerves.

The scrotal extension of the original peritoneal diverticulum communicates for a time with the rest of the celom through the processus vaginalis peritonei. The processus, however, eventually degenerates and persists only as a ligamentous thread inside the spermatic cord. The now isolated pocket of peritoneal cavity lining the scrotum is the **tunica vaginalis testis** (Fig. 22.3 c). The testis, which lies behind it, therefore is covered by tunica vaginalis in front and on its sides only. Occasionally, the processus fails to obliterate, in which case abdominal contents is able to herniate into the spermatic cord.

The Inguinal Canal

The testis passes through the developing tissues of the fetal abdominal wall early in the seventh month of pregnancy and arrives in the scrotum by the end of the eight month, with the spermatic cord in tow. It leaves behind a trail of passage, which persists in postnatal life as the inguinal canal. The canal has recognizable anterior and posterior walls, a roof, and a floor. The floor is ligamentous; the roof and walls are both muscular and fibrous.

The aponeurosis of the external oblique muscle makes important contributions to the inguinal canal. I have already referred to its free inferior border, the inguinal ligament—also known as Poupart's ligament. Its split attachment medially to both the pubic tubercle and the medial part of the pubic crest creates the **crura** (L., *crus* = a leg) of the superficial ring (Fig. 22.4). Extending laterally from the pubic tubercle, and bound to the pectineal line on the superior pubic ramus, is the strongly fibrous **pectineal ligament** (of Cooper). On its medial side is the **lacunar ligament** (of Gimbernat), a flat triangular membrane. It stretches horizontally between the medial ends of inguinal and pectineal ligaments, and forms the floor of the inguinal canal medially. Laterally, the inguinal ligament, whose unattached border folds to the inside, creates additional floor space.

The inguinal ligament is a key element in the construction of the inguinal canal for another reason. Both transversus abdominis and internal oblique muscles arise, in part, from it: transversus from the lateral one-third of the ligament, and internal oblique from the lateral two-thirds (Fig. 22.5 a). From their ligamentous origin above and in front of the spermatic cord, the muscles sweep across the cord and, as **conjoint tendon** insert below into the pubic crest behind the spermatic cord. The arching, spiralling fibers of transversus abdominis and internal oblique, therefore, form the roof of the inguinal canal. In addition, the muscles become portions of the anterior wall laterally and of the posterior wall medially. Transversalis fascia adds to the posterior wall of the inguinal canal laterally, and the aponeurosis of the external oblique completes the anterior wall medially (Fig. 22.5 b).

It follows, then, that fleshy fibers of internal oblique and transversus abdominis muscles are in front of the deep inguinal ring while the aponeurotic conjoint tendon is in back of the superficial inguinal ring. But, because the internal oblique has a more extensive origin from the inguinal ligament than does transversus, it forms a much greater part of the anterior wall of the inguinal canal.

Conversely, transversus abdominis, making a greater contribution to the conjoint tendon than does internal oblique, forms more of the posterior wall of the canal.

Inguinal Hernias

As the fetal testis began its extraperitoneal descent through the developing abdominal wall, its steady progress produced a sleeve of transversalis fascia whose mouth is the deep ring—the entrance into the inguinal canal. It is the site where testicular vessels, ductus deferens and artery, cremasteric vessels and nerve (the genital branch of the genitofemoral nerve), *etc*., assemble before becoming part of the spermatic cord. With closure of the processus vaginalis peritonei, a shallow fossa of parietal peritoneum fills the deep ring.

Herniations into the inguinal canal are not uncommon, especially in the male. The inguinal region is less of a problem area in the female, presumably because the ovaries do not enter the inguinal canal at any time and, thus, never stretch it. It is occupied, however, in fetal life by the **canal of Nuck**, the homologue of the male processus vaginalis peritonei; and, at all times, by the round ligament of the uterus, originally gubernaculum.

An inguinal hernia present at birth is a *congenital* inguinal hernia; it usually can be traced to a patent processus vaginalis. Hernias like it, that enter the canal via the deep ring, are **indirect** or **oblique** inguinal hernias. They occur postnatally when, for instance, increased intra-abdominal pressure, accompanying the lifting of heavy objects with the glottis closed, forces abdominal contents and parietal peritoneum past the deep ring into the canal. Sometimes, such an acquired indirect inguial hernia reopens the processus vaginalis peritonei, in which case the hernial sac has all the coverings of the spermatic cord; it may emerge at the superficial ring inside the cord, even descend into the scrotum.

A **direct** inguinal hernia enters the inguinal canal not through the deep ring, but by pushing up against the posterior wall of the canal—carrying parietal peritoneum, extraperitoneal tissue, and transversalis fascia into the canal ahead of it. As it enters the lower end of the inguinal canal, the hernial sac encounters conjoint tendon which may arrest its descent. Failing that, conjoint tendon rather than cremaster muscle becomes the outer investment of the hernial sac, which may exit at the superficial ring.

Indirect hernias may be congenital or acquired; direct inguinal hernias are always acquired. Indirect hernias may invade the spermatic cord; direct inguinal hernas are never inside it. Indirect hernias outnumber direct hernias by a ratio of 3:1.

Direct and indirect inguinal hernias also differ with respect to the location of the neck of the hernial sac—*i.e.*, the constricted part of the hernia projecting into the canal. The neck of the sac of a direct hernia lies within the **inguinal**

(Hesselbach's) **triangle**. It is located, therefore, *medial* to the inferior epigastric artery which, slanting upwards towards rectus abdominis, forms the lateral side of the triangle (Fig. 22.6). The medial side of the triangle is formed by the lateral border of the rectus muscle, and its base is the inguinal ligament. The floor of the inguinal triangle is the posterior wall of the inguinal canal. In contrast, the neck of an indirect hernial sac lies *lateral* to the inferior epigastric artery, outside the inguinal triangle, at the site of the deep inguinal ring.

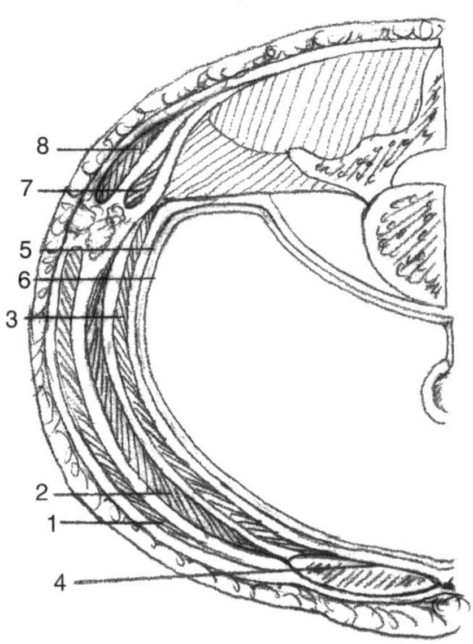

1. External oblique muscle
2. Internal oblique muscle
3. Transversus abdominis muscle
4. Rectus sheath and rectus abdominis muscle
5. Transversalis fascia
6. Parietal peritoneum
7. Serratus posterior inferior muscle
8. Latissimus dorsi muscle

22.1 Hemi-section of anterolateral abdominal wall

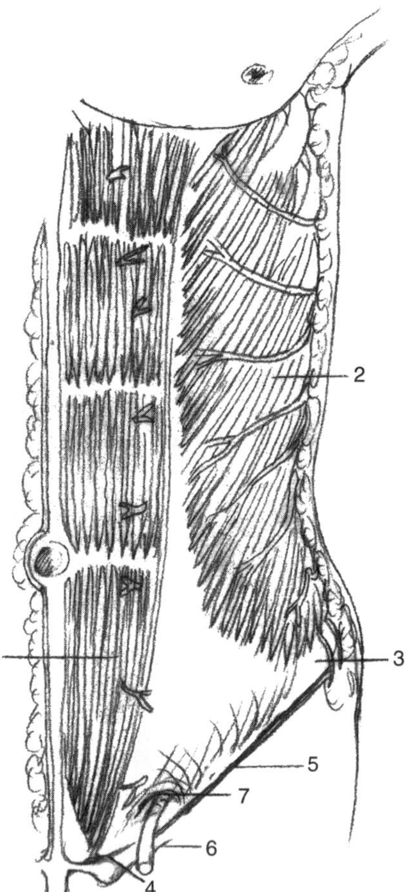

1. Rectus abdominis m.
2. External oblique m.
3. Anterior superior iliac spine (ASIS)
4. Pubic crest
5. Inguinal ligament
6. Spermatic cord (Round ligament of uterus)
7. Superficial inguinal ring

22.2 Left inguinal region

Figure 22.1: Cross-section of hemi-abdominal wall
Figure 22.2: Ventral view of hemi-abdominal wall (after Maxwell)

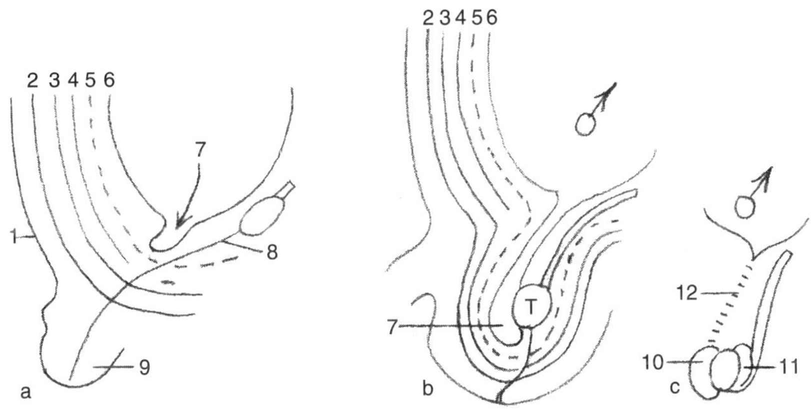

1. Skin and subcutaneous tissue
2. External oblique muscle
3. Internal oblique muscle
4. Transversus abdominis muscle
5. Transversalis fascia
6. Parietal peritoneum
7. Processus vaginalis peritonei
8. Fetal testis with gubernaculum
9. Scrotum
10. Tunica vaginalis testis
11. Epididymis and ductus deferens
12. Processus vaginalis closed

22.3 a-c Descent of fetal testis

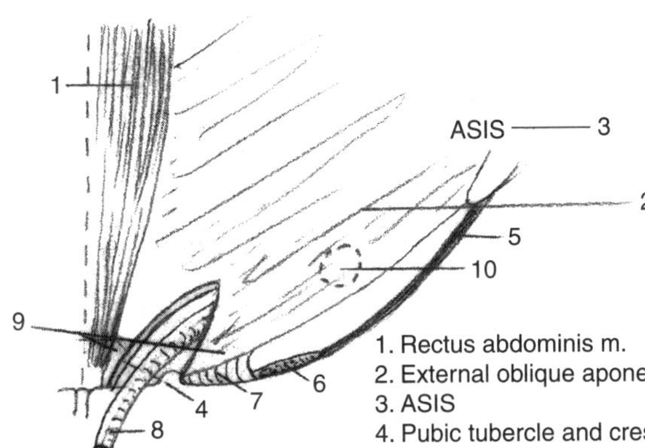

1. Rectus abdominis m.
2. External oblique aponeurosis
3. ASIS
4. Pubic tubercle and crest
5. Inguinal ligament
6. Pectineal ligament
7. Lacunar ligament
8. Spermatic cord (Round ligament of uterus)
9. Crura of superficial inguinal ring
10. Site of deep inguinal ring

22.4 Ligaments of inguinal region

Figure 22.3 a-c: Descent of fetal testis
Figure 22.4: Ligaments of inguinal region

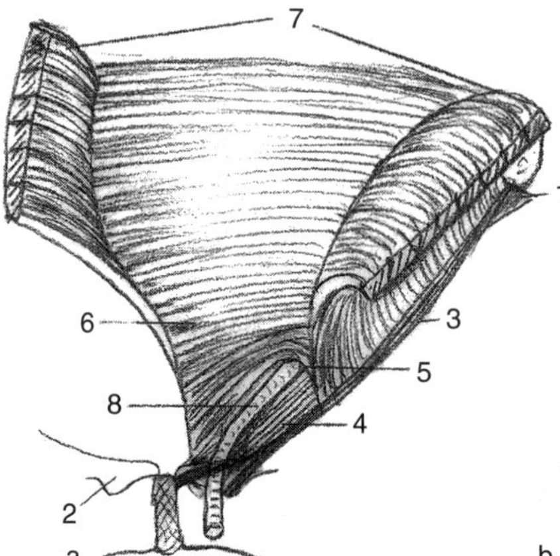

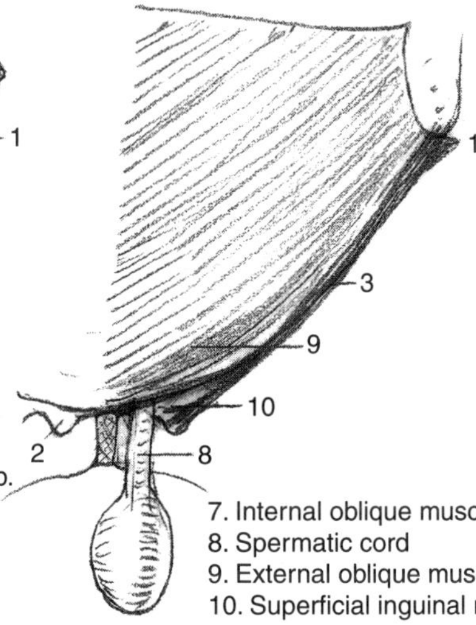

1. ASIS
2. Pubic tubercle and crest
3. Inguinal ligament
4. Transversalis fascia
5. Deep inguinal ring
6. Transversus abdominis muscle
7. Internal oblique muscle
8. Spermatic cord
9. External oblique muscle
10. Superficial inguinal ring

22.5 a-b Inguinal canal

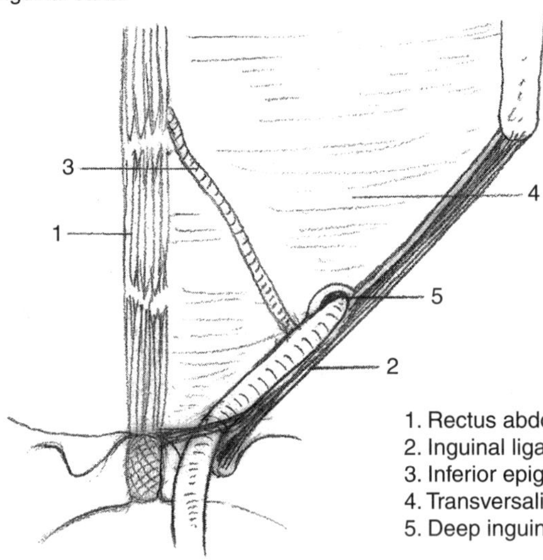

1. Rectus abdominis muscle
2. Inguinal ligament
3. Inferior epigastric artery
4. Transversalis fascia
5. Deep inguinal ring

22.6 Inguinal (Hesselbach's) triangle

Figure 22.5 a-b: Inguinal canal
Figure 22.6: Inguinal (Hesselbach's) triangle

Chapter 23
DEVELOPMENT OF THE DIGESTIVE SYSTEM

The adult digestive system is the direct descendant of the primitive yolk sac; its germ layer source, therefore, is embryonic endoderm. Towards the end of the third week of human development, head and tailfolds undercut the yolk sac, so that three distinct parts of the digestive system can be recognized; they are (1) foregut, (2) midgut, and (3) hindgut (Fig. 23.1). At first a straight midline tube between future mouth and anus, the embryonic digestive system is suspended from the dorsal body wall by the **dorsal mesentery**. This double-layered membrane transmits the doral aorta and its three unpaired, abdominal branches: celiac, superior and inferior mesenteric arteries—supplying, respectively, foregut, midgut, and hindgut. A **ventral mesentery** connects ventral body wall and foregut. Its lower, free, sickle-shaped border, the **falciform ligament**, transmits the umbilical vein, passing from the placenta through the umbilical ring towards the sinus venosus. To get there, the umbilical vein must navigate the septum transversum.

The Convoluted Stage

In embryos of six weeks, some growth of the digestive tract has taken place. The tube is now longer than the dorsal mesentery and that part of the body wall along which it runs. It turns, twists, and soon deviates from the midline in several places, forming in the process the prominent duodenal and midgut loops. Figure 23.1 illustrates the basic plan and outcome of its parts: the foregut in the headfold gives rise to the pharynx, esophagus, stomach, and the proximal one-half of duodenum, as well as to both respiratory and hepatic diverticula. The midgut develops into small intestine, starting with the distal one-half of duodenum, and large intestine up to, and including, the proximal two-thirds of transverse colon. The hindgut in the tailfold turns into the rest of the large bowel.

The apex of the **duodenal loop** points to the right and is held in place by the falciform ligament (Fig. 23.2). The attachment of the ligament to the gut wall marks the transition between foregut and midgut regions. The proximal limb of the duodenal loop, therefore, is foregut territory; the distal limb, midgut.

The **midgut loop** lies in the median plane and, for the time being, curves forward (Fig. 23.1). Its apex is continuous distally through the umbilical ring with the vitello-intestinal duct and degenerating yolk sac. The roots of the loop are firmly secured to the dorsal body wall by condensations in the dorsal mesentery—the so-called retention bands. The proximal limb of the midgut loop develops into the jejunum and all but the last few coils of ileum; the distal limb forms the rest of the ileum, cecum, appendix, ascending colon, and transverse colon, except for its terminal one-third.

Focus on the Foregut

By the end of the fourth week of human development the oropharyngeal membrane has disintegrated, continuity between oral cavity and primitive pharynx has been achieved, and amniotic fluid swallowed by the embryo passes into the foregut.

I traced the early development of the primitive pharynx in chapter 12, but must return to it now, for much important detail was left unsaid at that time.

The embryonic pharynx is the direct source of the *parenchyma* of several major organs—*i.e.*, of cell populations carrying out organ-specific functions. The lungs and thyroid gland both begin as outgrowths from the floor of the primitive pharynx—the former as the **pulmonary diverticulum**, the latter as the **thyroglossal duct**. The alveolar epithelial cells of the lung and the secretory cells of the thyroid gland, therefore, are endodermal in origin. (*N.B.*, the non-specific components, *e.g.*, connective tissue, form the *stroma* of organs.)

The pulmonary outgrowth begins as the laryngotracheal groove, a keyhole shaped opening in the floor of the pharynx (Fig. 23.3). While retaining its connection to pharynx, the diverticulum elongates and descends in front of the pharynx for a short distance, then divides to form right and left lung buds. The initial segment of the pulmonary outgrowth persists as the mucosal epithelium of larynx and trachea. Repeated branching of the lung buds creates air-conducting passages of ever diminishing size leading to thin-walled air sacs (alveoli) where physiologic exchange takes place.

The thyroid gland develops as a median endodermal thickening in the floor of the primitive pharynx over the second pharygeal arch. The thickening develops into a hollow passage, the thyroglossal duct, which grows tailwards ventral to the budding respiratory system (Fig. 23.3). The duct bifurcates near the commencement of the trachea to form the isthmus and lobes of the thyroid gland, then usually disappears without a trace. The foramen cecum (Fig. 23.4), located at the apex of the sulcus terminalis dividing body and root of the tongue, marks the site of the original thyroglossal outgrowth.

And, speaking of the **tongue**, the mucosal epithelium of its root (posterior one-third) is believed to come from the *copula*—the upper part of the hypobranchial eminence, a mass of endoderm in the floor of the primitive pharynx between thyroid and pulmonary rudiments. The lower part of the hypobranchial endoderm becomes the epithelial lining of the epiglottis. The mucosal epithelium of the tongue's body (anterior two-thirds) probably arises by the fusion of two distal tongue buds with the tuberculum impar (proximal median tongue bud). The germ layer source of this epithelial layer is still in doubt, though it is generally assumed that the cells just in front of the sulcus terminalis are derived from foregut endoderm. Until we know the exact position and attachments of the oropharyngeal membrane in the early embryo, however, the precise territorial boundary line between stomodeum (ectoderm) and foregut (endoderm) will remain elusive.

The Hepatopancreatic Diverticulum

The functional cell populations of the liver, gall bladder, and pancreas all arise from the **hepatopancreactic diverticulum**, an endodermal outgrowth from the apex of the duodenal loop. As it invades the mesoderm of the ventral mesentery, the diverticulum distally divides in the substance of the septum transversum into two stems (Fig. 23.1, 23.2 a). Their tips bud off epithelial cells, the future hepatocytes, which colonize the septum transversum. Septal tissue, therefore, becomes the stroma to the liver's parenchyma.

Just proximal to the distal split, an additional stem and bud mark the primordium of the **gall bladder** and **cystic duct**. More proximal still, is the ventral pancreatic rudiment and its duct.

The initial segment of the hepatic outgrowth remains as the **common bile duct**; the rest, beyond the entrance of the cystic duct, persists as the **common hepatic duct** receiving right and left hepatic ducts. When organogenesis is complete, the common bile duct opens, as you might expect based on its developmental history, into the middle section of the descending part of the duodenum. Before terminating on the major duodenal papilla (Fig. 23.2 b), the common bile duct is joined by the main duct of the pancreas to form the **hepatopancreatic ampulla** (of Vater).

The **pancreas** has ventral and dorsal primordia. The **ventral pancreatic rudiment**, as just noted, is an outgrowth from the hepatic diverticulum near its origin. The **dorsal pancreatic rudiment**, however, arises from the concavity of the duodenal loop. As the apex of the loop rotates, presumably because the right (convex) border outgrows the left (concave) border, the common bile duct is shifted to the left and to the back of the duodenum (Fig. 23.2 b). This move brings the two primordia into close apposition and causes them to fuse. The ventral rudiment becomes the head and neck of the future pancreas, the dorsal rudiment its body and tail. The two ducts systems merge in such a manner that the **pancreatic duct** (of Wirsung) drains all of the ventral rudiment plus the distal part of the dorsal rudiment. The proximal portion of the dorsal duct remains as the **accessory pancreatic duct** (of Santorini); it opens independently into the descending part of the duodenum.

The Midgut Loop

Towards the sixth week of human development, the midgut loop rotates some 270° in a counter-clockwise direction, using the superior mesenteric artery as axis of rotation (Fig. 23.5 a–e). Once the process has run its course, the two limbs no longer occupy the median plane but lie side by side—the proximal limb on the left and the distal limb on the right. Rapid elongation of the loop, especially the part fated to become small intestine, causes it to pass through the umbilical ring into the umbilical cord. The herniation of the midgut loop to a location outside the body at this stage of development is normal. Furthermore, it is only temporary. Some four weeks after leaving the body's interior, as the abdominal cavity grows roomier, the loop re-enters the abdomen. The proximal limb, *i.e.*, future jejuno-ileum, returns first and, because the bulky liver fills most of the available space on the right, lays claim to the left side of the abdominal cavity; its highly convoluted coils push the future descending colon (hindgut) towards the left edge. The distal limb is subsequently led back into the body and gains access to right side of the cavity. These facts explain why terminal ileum, cecum with appendix, and ascending colon occupy right-lateral positions in the abdomen; they also account for the right-to-left cross-over of the transverse colon in front of the duodenum (Fig. 23.5 e).

Addenda

A few additional comments concerning the lower digestive tract, and a short summary of germ layer sources may be relevant at this point.

The junction between the proximal two-thirds and distal one-third of the transverse colon indicates not only the site where midgut turns into hindgut. It is also the zone separating the bloodvascular domains of superior mesenteric (midgut) and inferior mesenteric (hindgut) arteries and veins, and the region beyond which the vagus nerves do not distribute.

As development continues, the colon increases in width. The **transverse colon** also increases in length and begins to sag, foreshadowing the appearance of **right colic** (hepatic) and **left colic** (splenic) **flexures**. In general, development of the hindgut, *e.g.*, descending and sigmoid colons, lags behind that of the rest of the digestive tube. All parts of the large bowel *except* for the rectum, however, eventually acquire **taeniae coli** (banded, longitudinal smooth muscle), **omental appendices** (tabs of fat), and **haustra** (pleats).

The anatomy of the anorectal segment will be discussed in subsequent chapters.

The following important message bears repeating: *the mucosal epithelium of the digestive system and its outgrowths between oropharyngeal and cloacal membranes is derived exclusively from yolk sac endoderm.* As discussed here, the derivatives include the parenchyma of lung, liver, and gall bladder as well as the secretory cells of the thyroid, pancreas, and gastric and intestinal glands. And, equally important, *mucosal epithelia and diverticula that are outside oropharyngeal and cloacal membranes come from embryonic ectoderm.* The epithelial outgrowths of the oral pit (stomodeum) include tooth buds as well as Rathke's pouch, the source of adenohypophysial secretory tissue. Ectoderm-derived epithelium lines the lower one-third of the anal canal (proctodeum). The germ layer source of the connective tissue framework that supports organic-specific cells is, of course, always mesoderm.

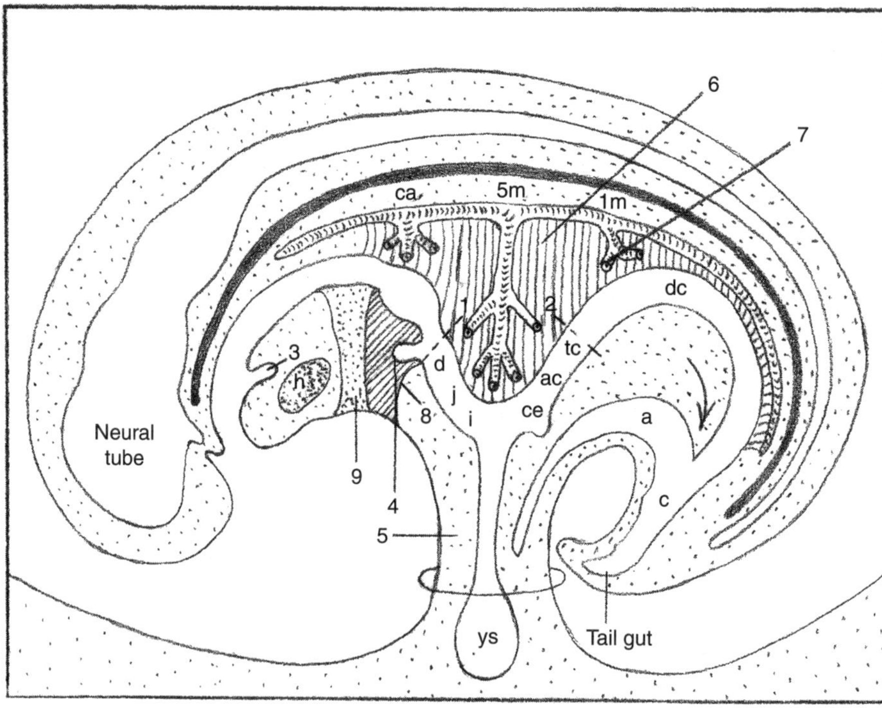

1. Divisions between foregut/midgut
2. Division between midgut/hindgut
3. Pulmonary diverticulum
4. Hepatopancreatic diverticulum
5. Umbilical cord
6. Dorsal mesentery
7. Branch of inferior mesenteric a.
8. Falciform margin of ventral mesentery
9. Septum transversum

a. Allantois
c. Cloaca

23.1 Divisions of primitive alimentary canal

Figure 23.1: Divisions of the primitive alimentary canal

23.2 a-b Duodenal loop and outgrowth

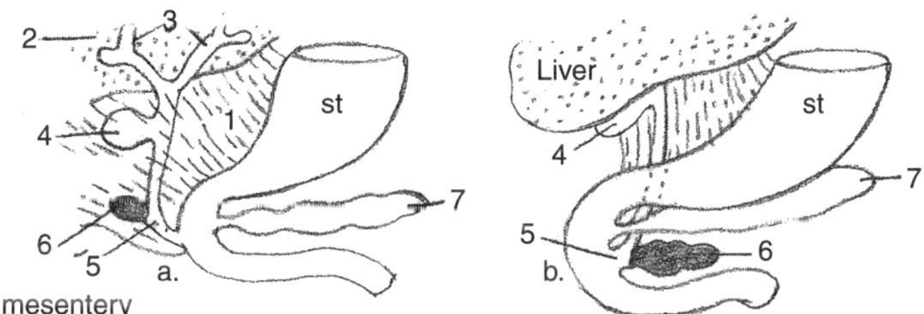

1. Ventral mesentery
2. Septum transversum
3. Hepatic ducts
4. Gall bladder

5. Common bile duct
6. Ventral pancreatic rudiment
7. Dorsal pancreatic rudiment

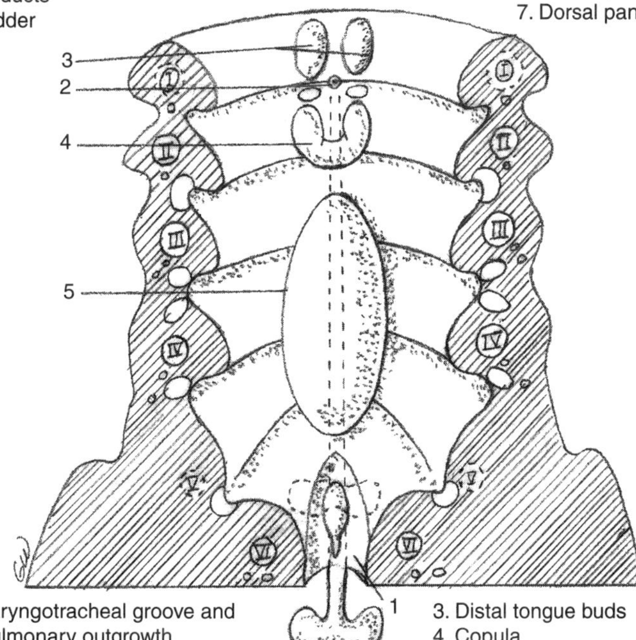

1. Laryngotracheal groove and pulmonary outgrowth
2. Outgrowth of thyroglossal duct
3. Distal tongue buds
4. Copula
5. Hypobranchial eminence

23.3 Floor of primitive pharynx

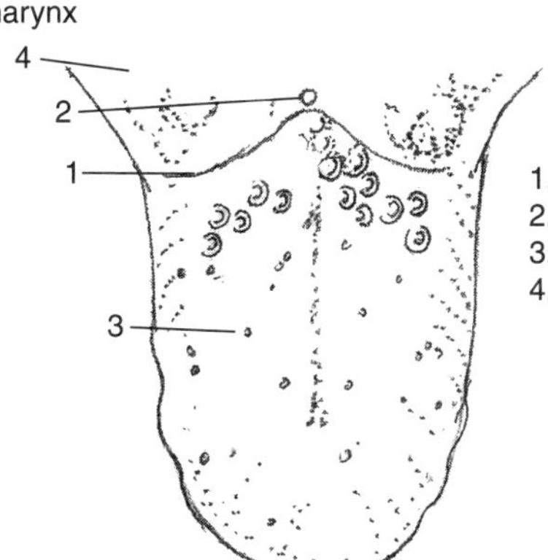

1. Sulcus terminalis
2. Foramen cecum
3. Body of tongue
4. Root of tongue

23.4 The tongue

Figure 23.2 a-b: The duodenal loop and outgrowth
Figure 23.3: Floor of the primitive pharynx
Figure 23.4: The tongue

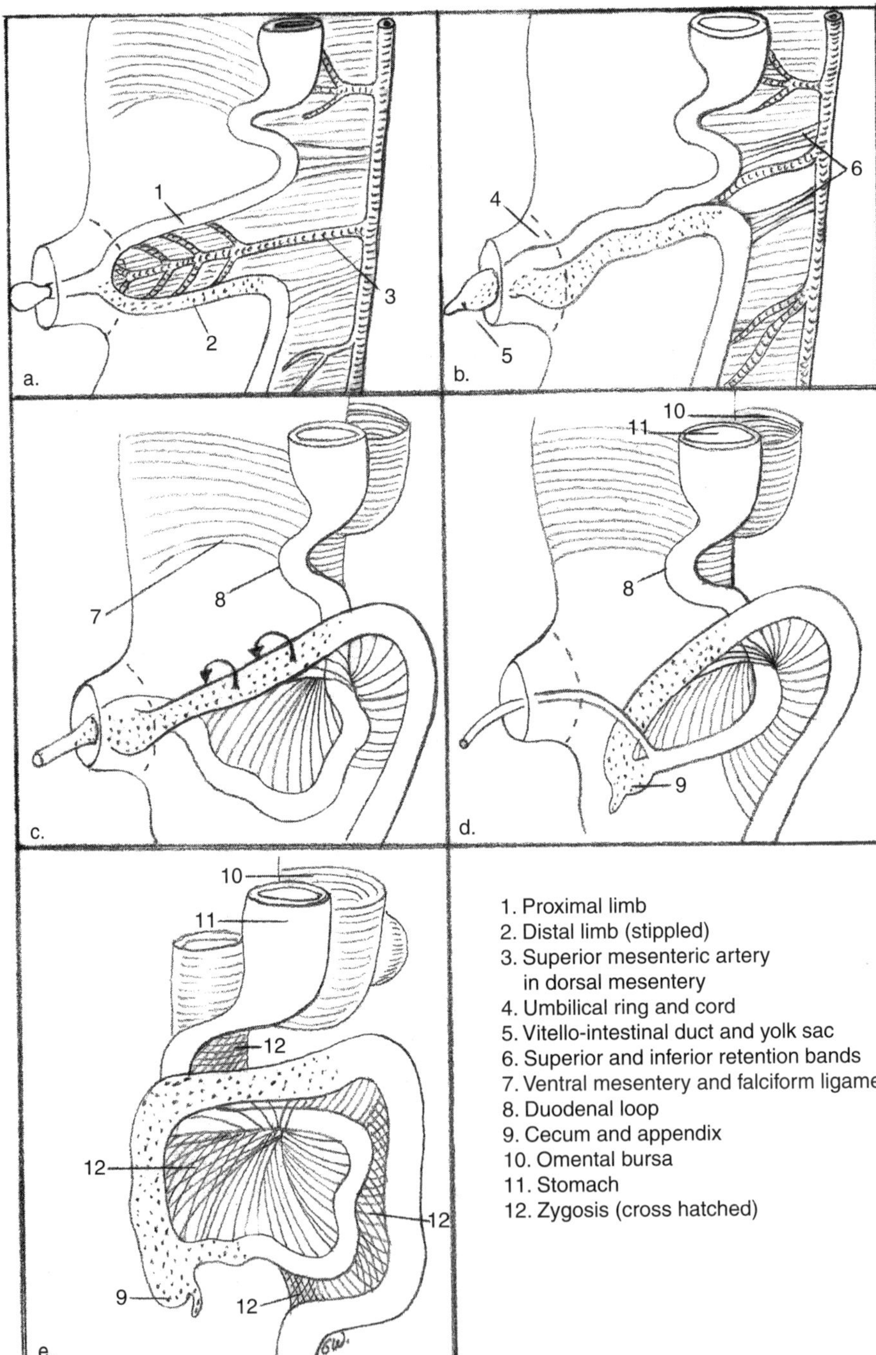

1. Proximal limb
2. Distal limb (stippled)
3. Superior mesenteric artery in dorsal mesentery
4. Umbilical ring and cord
5. Vitello-intestinal duct and yolk sac
6. Superior and inferior retention bands
7. Ventral mesentery and falciform ligament
8. Duodenal loop
9. Cecum and appendix
10. Omental bursa
11. Stomach
12. Zygosis (cross hatched)

23.5 a-e Rotation of midgut loop

Figure 23.5 a-e: Rotation of midgut loop

Chapter 24
THE PERITONEUM

The peritoneum, divisible into visceral and parietal layers, is the serous membrane that "wall-papers" the peritoneal cavity. **Visceral** peritoneum forms the outer coat (serosa) of hollow abdominal organs and can be separated from them only with difficulty. **Parietal** peritoneum is firmly applied to the body wall internally. In a number of places, parietal peritoneum is reflected off the body wall to become continuous with visceral peritoneum—in the same way that parietal pericardium is continuous with visceral pericardium (epicardium) of the heart. The double-layered folds of peritoneum connecting parietal peritoneum of body wall and visceral peritoneum over internal organs are mesenteries.

Like other body cavities, *e.g.*, pericardial and pleural, the peritoneal cavity is a closed space containing a small quantity of watery (serous) fluid. Formation of this cavity is closely tied to the establishment of the ventral body wall in the embryo. Here is a brief summary of its developmental history.

Creation of the Ventral Body Wall

Cross-sectional views of the human embryo (Figs. 24,1 a–b) illustrate lateral folding during the fourth week of intra-uterine life. Right and left lateral body folds grow ventrally, meet in front, and fuse in the midline—thus establishing the ventral (anterior) body wall and converting a flat embryonic plate into a cylindrical embryo. It is, of course, the amniotic cavity expanding in a ventral direction that initiates lateral folding (think of it as the embryo's attempt to wrap itself in a blanket of amniotic fluid); and, when the midventral fusion membrane breaks down, the embryo lies free inside the amniotic cavity. An opening in the ventral body wall, the umbilical ring, admits the vitello-intestinal duct and lets umbilical vessels pass to and from the placenta.

Lateral folding, coincidentally, causes the roof of the yolk sac to roll up into a tube and to be internalized as the primitive digestive system. Coincidentally also, establishment of the ventral body wall creates the embryonic celom. Its wall is lined internally with lateral plate (somatopleuric) mesoderm, forming the striated musculature of the anterolateral abdominal wall, fascia transversalis, and parietal peritoneum. The lateral plate (splanchnopleuric) mesoderm applied to yolk sac endoderm, develops into the smooth muscle of the digestive tube, connective tissue, and visceral peritoneum. The primitive celom becomes the **peritoneal cavity**.

Be sure you understand the difference between abdominal cavity and peritoneal cavity. The *abdominal cavity* is the body cavity below the diaphragm. It is divisible into the abdominal cavity proper and pelvic cavity, and is composed of bones, cartilage, muscle, aponeuroses, and fascia. The abdominal cavity contains both digestive and urogenital organs, their vessels and nerves. The *peritoneal cavity* is an closed space within the abdominal cavity, lined by a single row of mesothelial cells. Its sole constituent is serous fluid.

Description of Peritoneum

The arrangement of the peritoneum in the adult is complicated, because many internal organs change form and position during development. Growth and elongation of the tubular digestive tract causes it to twist and turn, which accounts for its deviation from the midline and coiled appearance. And, as parts of the gut flex and loop, the dorsal and ventral mesenteries attaching the system to the body wall also become displaced and twisted, causing them to be thrown into folds and ruffles. These developmental changes affecting gut and mesenteries reduce the original roomy dimensions of the peritoneal cavity to a narrow, convoluted, meandering, "potential" space in the fully formed body.

To help you visualize the events that influence the ultimate shape of the peritoneal cavity, we'll review the deployment of the digestive tract in a four-week-old human embryo as seen in cross-section (Fig. 24.2 a–c). That part of the primitive digestive tube destined to become stomach has expanded in a dorsoventral direction and is suspended from both dorsal and ventral body walls by dorsal and ventral mesogastria, respectively (Fig. 24.2 a). As the stomach rotates clockwise some 90° (Fig. 24.2 b), its right surface turns towards the back and its dorsal border—the future greater curvature—points to the left. The left side of the stomach now faces forward and its ventral border, becoming lesser curvature, is on the right.

Concomitant with stomach rotation, the dorsal mesogastrium balloons to the left. In its substance, aggregations of mesoderm coalesce to form the **spleen** (lien). A blood-forming organ in the embryo, the spleen later becomes part of the body's defense system concerned with lymphocyte and antibody production. The enlarging spleen divides the dorsal mesogastrium into **gastrosplenic** and **splenorenal ligaments**. The ventral mesogastrium swings to the right. It is split by the liver, developing in its substance, into the **lesser omentum** (hepatogastric plus hepatoduodenal ligaments) and the **falciform ligaments**. The **round ligament of the liver**, representing the obliterated fetal left umbilical vein occupies the lower free border of the falciform ligament.

Ligaments and omenta, you will recall, are double-layered peritoneal folds. In general, mesenteries attach organs to the body wall; ligaments and omenta connect one organ with another.

Rearrangement of dorsal and ventral mesenteries also changes the topography of the peritoneal cavity in the foregut region: at first lying side by side, the two parts of the peritoneal cavity now become transversely oriented. The part of the cavity trapped between stomach and dorsal body wall is the **omental bursa** (lesser sac of peritoneum). It communicates with the rest of the peritoneal cavity (greater sac) through the **omental foramen** (of Winslow).

Zygosis

Mesenteries and similar peritoneal folds are double-layered membranes of mesothelium with a core of extraperitoneal connective tissue through which vessels and nerves reach organs they supply. It is not uncommon for mesothelial cells of membranes separated only by a thin layer of fluid to be worn away when subjected to constant pressure or friction for extended periods of time, as happens during normal embryonic development. The exposed extraperitoneal connective tissue, then, characteristically, fuses across the peritoneal space and eliminates it (Fig. 24.3 a–b). The process by which such fusion zones, literally *fusion fascias*, are established is known as **zygosis** (Gr., *zygo* = a junction). Zygosis is the mechanism by which organs suspended by mesenteries become immobilized and fixed to the body wall, causing them to become *retroperitoneal.* A specific example may help clarify the concept: When the mesentery of the fetal duodenum, the mesoduodenum, is jammed up against the dorsal body wall following the reduction of the (physiological) umbilical hernia (see chapter 23), mesothelial cells are lost from both mesoduodenum and parietal peritoneum. The narrow peritoneal space formerly separating the two layers is abolished, a fusion fascia forms, and the duodenum, including its outgrowth—the pancreas, are rendered retroperitoneal. A similar fate befalls both ascending and descending colons following the 270° anti-clockwise twist of the dorsal mesentery (see Fig. 23.5, p. 166). The rectum, likewise, fails to retain its mesorectum. The term *retroperitoneal* implies, of course, that only front and sides of these organs are covered by peritoneum.

Other retroperitoneal organs, such as the kidneys or adrenal glands, owe their location not to zygosis-related events, but to the manner in which they developed.

Fusion fascias are useful to surgeons who wish to mobilize an organ such as the duodenum without endangering its blood or nerve supply; for, cleavage in the plane of fusion simply re-establishes the original mesentery.

While certain parts of the primitive dorsal mesentery are lost to zygosis, other segments persist. They include *the* **mesentery**, attaching the length of jejuno-ileum to the dorsal body wall, the **transverse mesocolon** suspending the transverse colon, and the **sigmoid mesocolon**.

The Greater Omentum

The **greater omentum** arises as a redundancy of the primitive dorsal mesogastrium after the dorsal border of the stomach, the future greater curvature, swings to the left and forward. The greater omentum is frequently the first structure you encounter upon opening the peritoneal cavity. It is a double walled pouch of peritoneum containing the lower recess of the omental bursa. Layers 1 and 2 of the greater omentum form the anterior wall of the recess and layers 3 and 4 its posterior wall (Fig 24.4 a–b).

Downgrowth of the greater omentum brings its undersurface into contact with the upper surfaces of transverse colon and mesocolon. Mesothelial cells of apposing membranes are worn away, causing fusion fascias to develop between them. Thereafter, the greater omentum connects the greater curvature of the stomach with the inferior border of the transverse colon; hence, its alternative name—gastrocolic ligament. Suspended like an apron, the greater omentum conceals many of the organs located in the epigastric region—including the transverse colon.

In obese individuals, the greater omentum is laden with fat. It helps protect abdominal contents from physical injury and also from bacterial infections, for it contains vast numbers of lymphocytes. The greater omentum has the capacity of wrapping itself around a locus of infection, thus isolating the site from the rest of the abdominal cavity. So much so, that some have dubbed it the "Policeman of the Abdomen."

The Omental Bursa

This anatomical cul-de-sac has been likened to a hot-water bottle lying sideways, its mouth pointing to the right (Fig. 24.2 c). The boundaries of the omental bursa are as follows (Figs 24.2, 24.4): in front, the posterior surface of the stomach and, at the far left, the pedicle of the spleen—*i.e.*, gastrosplenic and splenorenal ligaments; behind, parietal peritoneum over pancreas, upper left kidney, and left adrenal gland. The roof of the omental bursa is formed by the diaphragm and includes the upper recess between inferior vena cava and esophagus; the caudate lobe of the liver forms the anterior wall of this recess. The floor is continuous with the inferior recess between anterior and posterior layers of the greater omentum. Commonly, however, the lower recess of the omental bursa is absent, because layers 2 and 3 of the greater omentum have lost their mesothelial linings and fused. The omental foramen at the extreme right, at first open-mouthed, becomes greatly constricted as the liver overgrows it. Its margins are: the free border of the lesser omentum in front, the inferior vena cava covered by parietal peritoneum behind, the liver above, and the superior part of the duodenum below.

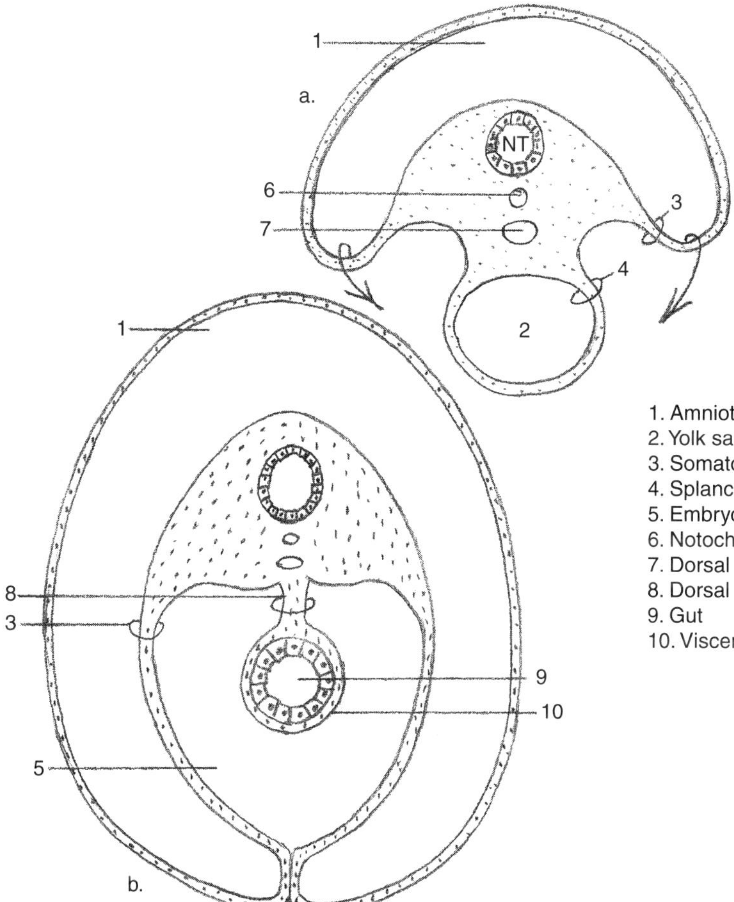

1. Amniotic cavity
2. Yolk sac
3. Somatopleure
4. Splanchnopleure
5. Embryonic celom
6. Notochord
7. Dorsal aorta
8. Dorsal mesentery
9. Gut
10. Visceral peritoneum

24.1 a-b Development of ventral body wall

Figure 24.1 a-b: Development of ventral body wall

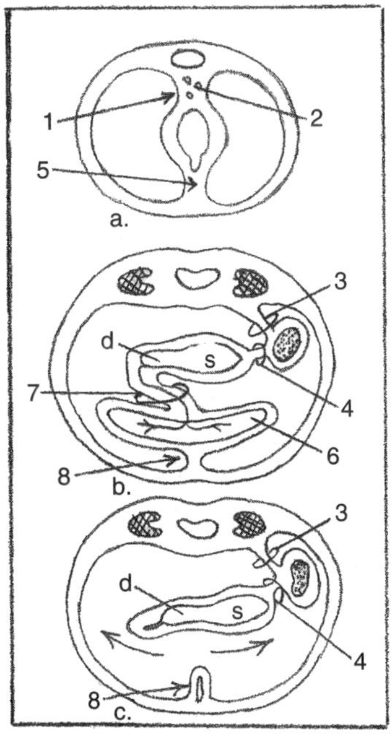

1. Dorsal mesogastrium
2. Developing spleen
3. Splenorenal ligament
4. Gastrosplenic ligament
5. Ventral mesogastrium
6. Developing liver
7. Lesser omentum
8. Falciform ligament
d. Duodenum
s. Stomach

24.2 a-c Abdominal cavity
(sequential stages in cross-sections)

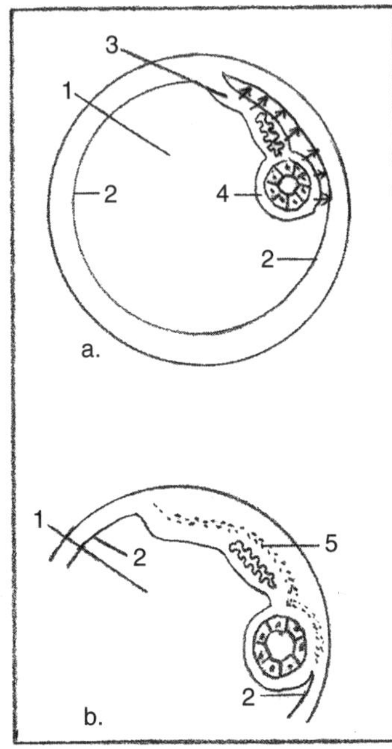

1. Peritoneal cavity
2. Parietal peritoneum
3. Mesoduodenum
4. Visceral peritoneum
5. Fusion fascia

24.3 a-b Zygosis a. Before b. After

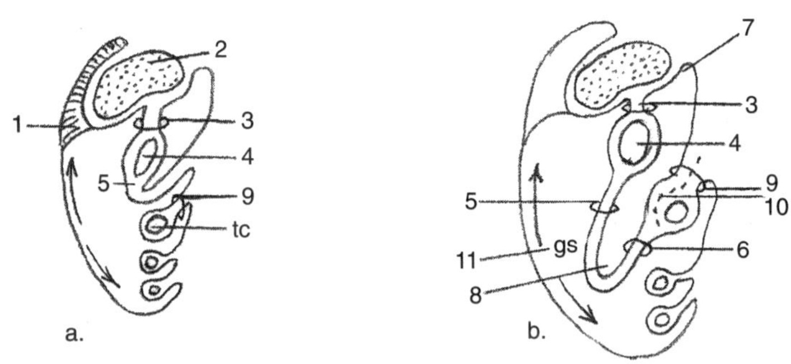

1. Falciform ligament
2. Liver
3. Lesser omentum
4. Stomach
5. Anterior wall of greater omentum and omental bursa
6. Posterior wall of greater omentum and omental bursa
7. Superior recess of omental bursa
8. Inferior recess of omental bursa
9. Transverse mesocolon
10. Fusion fascia
11. Peritoneal cavity (Greater sac)

24.4 a-b Greater omentum (sequential stages in vertical sections)

Figure 24.2 a-c: Abdominal cavity (cross-sections)
Figure 24.3 a-b: Zygosis a. Before b. After
Figure 24.4 a-b: Greater omentum (vertical sections)

Chapter 25
ABDOMINOPELVIC CONTENTS

The (respiratory) diaphragm forms the roof of the abdominopelvic cavity; the pelvic diaphragm, separating the cavity from the perineum, is its floor. The diaphragm arises from the xiphoid process and costal margin, and by right and left crura, from the bodies of the upper three lumbar vertebrae. The pelvic diaphragm stretches across the pelvic outlet, framed by bones and ligaments.

The diaphragms are striated muscle, sandwiched between fascia. Both are dome-shaped, though they are concavoconvex in opposite directions; and both transmit structures entering or leaving the abdominopelvic cavity.

The topographic relationships of abdominal organs, like those of the thorax or the neck, are best expressed with reference to some imaginary plane and/or some relatively stable skeletal marker, such as a vertebral segment. The **transpyloric plane** qualifies as a useful paradigm, and I shall use it here to map abdominopelvic contents.

The Transpyloric Plane

This imaginary plane is indicated on the body surface by a horizontal line placed midway between the jugular (suprasternal) notch and symphysis pubis (Fig. 25.1). Anteriorly, the line intersects the middle of the costal margin where the ninth costal cartilage meets the vertical linea semilunaris (the lateral border of the rectus abdominis muscle); posteriorly, it passes over the 12th rib where the bone is crossed by the lateral edge of the sacrospinalis muscle.

The transpyloric plane cuts across **vertebra L1**—though, once again, be aware that vertebral levels in the erect as well as in the recumbent living body may vary by as much as one whole segment, depending on age, sex, posture, and respiratory movement. But, an impressive amount of anatomical detail clusters at this vertebral level; and, putting this landmark to use brings sense and order to the study of abdominal contents.

To begin with, both the **pylorus** of the stomach and the **superior part of the duodenum** are clearly identified with the transpyloric plane (Fig. 25.2). Just above the gastroduodenal junction is the omental foramen and the free margin of the lesser omentum containing common bile duct, proper hepatic artery, and portal vein.

The **pancreas** fills the concavity of the duodenum. Its body and tail pass to the left at the level of the transpyloric plane. On the far left, the tip of the tail impales the **hilum of the spleen**, whose lower border, therefore, also lies near the transpyloric plane.

Emerging between head and neck of the pancreas is the **superior mesenteric artery**; it arises from the descending aorta opposite vertebra L1. The celiac trunk, therefore, comes off the aorta approximately one vertebral segment higher, between the crura of the diaphragm which border the aortic hiatus.

The superior mesenteric vein, passing to the right of the artery, is joined by the splenic vein behind the neck of the pancreas to form the portal vein at the level of the transpyloric plane.

The hilum of the **left kidney** is somewhat above the plane and that of the **right kidney** is somewhat below it. You may infer from this information that a trio each of **renal pelvis, artery,** and **vein** straddles the transpyloric plane. You may deduce also that the adrenal and gonadal branches of the aorta do not arise at the level of the transpyloric plane, but a short distance from it. Corresponding veins accompany the arteries. Keep in mind, however, that the right adrenal and gonadal veins terminate in the inferior vena cava directly, while those on the left drain into the left renal vein. Remember that the left renal vein is longer than the right vein, that it crosses in front of the aorta and returns blood to the inferior vena cava from the left adrenal gland and gonad as well as from the kidney.

Both **colic flexures** encroach on the transpyloric plane. The right (hepatic) flexure is lower than the left (splenic) flexure, for it is prevented from rising higher—as is the right kidney—by the liver above. The **transverse mesocolon**, which suspends the transverse colon from the posterior body wall, arises in front of the pancreas—hence, at or about the level of vertebra L1.

Other structures being topographically related to the transpyloric plane are the beginning of the thoracic duct (cisterna chyli) and the roots of the azygos and (inferior) hemi-azygos veins.

Finally, the **fundus of the gall bladder**, situated deep to the right ninth costal cartilage near the middle of the costal margin, lies on the transpyloric plane.

You can solidify your knowledge of these important anatomical relationships by projecting onto the face of each kidney parts of organs in contact with it. As pictured in Fig 25.3, making impressions on the right kidney are the right lobe of the liver, the descending part of the duodenum, and the right colic flexure. Organs leaving an imprint on the left kidney are the stomach, the tail of the pancreas, and the left colic flexure. Adrenal glands and coils of small intestine fill in the rest.

The posterior relations of the kidney are illustrated in Fig. 25.4. The 12th rib divides the kidney roughly into upper two-thirds and a lower one-third (which, therefore, is not protected by bone). The diaphragm, as part of the posterior abdominal wall, is related to the back of each kidney superolaterally. Psoas, quadratus lumborum, and transversus abdominis muscles, in order, from medial to lateral, form the lower posterior relationships of the kidneys.

Other Transabdominal Planes

There are several other abdominal planes that are useful tools in the mapping of abdominopelvic contents (Fig. 25.1).

The **subcostal plane** is an imaginary transverse plane at the base of the costal margins. It intersects the spine near **vertebra L3** in quiet respiration and above L3 during deep inspiration. The subcostal plane indicates the position of the **inferior** (horizontal) **part of the duodenum** and the origin of the **inferior mesenteric artery**.

The **supracristal** (transumbilical) **plane** passes over the tops of the iliac crests and through the umbilicus. It is a guide to **vertebra L4**, to the **bifurcation of the abdominal aorta**, and the **beginning of the inferior vena cava**. Vertebral level L4, as you know, is a useful landmark in procedures, such as lumbar puncture, requiring access to the subarachnoid space.

The **interspinous plane** lines up at the level of the anterior superior iliac spines. It cuts across the **sacral promontory** and **vertebra S2**. The body's center of gravity is just anterior to, and opposite, the vertebra.

The Development and Anatomy of the Rectum

The expanded distal end of the embryonic hindgut is the **cloaca** (Fig. 25.5; see also Fig. 23.1, p. 164). It is closed off below by the **cloacal membrane**, which separates the cloaca from the **anal pit** (proctodeum). When the membrane ruptures, sometime during the seventh week of development, the cloaca opens into the amniotic cavity. Even before breakdown of the cloacal membrane, however, downgrowth of the urorectal septum partitions the cloaca into ventral and dorsal parts. The ventral part develops into **bladder** and **lower urogenital tract**; the dorsal part, which is continuous above with the sigmoid colon, becomes **rectum** and upper two-thirds of the **anal canal**. The lower one-third of the anal canal develops from the anal pit.

In the adult, the rectum is roughly 11.5 cm (4.5 in.) in length, and the anal canal measures 3.5 cm (1.5 in.). The pelvic diaphragm is the site of the **anorectal junction**; *i.e.,* the rectum is located in the pelvis, the anal canal is part of the perineum.

The lower end of the rectum is dilated to form the **ampulla**. Actually, the rectum is not straight, as its name implies. It curves twice in the median plane:

once, as it follows the concavity of the sacrum and coccyx (sacral flexure), and, again, as it passes out of the pelvis (Fig. 25.6). The apex of this right-angled anorectal flexure points in the direction of the umbilicus. Moreover, the rectum also has three lateral curvatures: two are concave to the left and one, the middle curvature, is concave to the right. They result because the longitudinal muscle present in the rectal wall is of shorter length than the rectum itself. The muscle, however, is no longer arranged in three bands (taeniae) so typical of the large intestine. In the rectal wall, longitudinal muscle forms a complete layer—thick in front and back, less so along the sides. The rectum differs from the colon also in lacking omental appendices.

The transition between sigmoid colon and rectum is opposite the third piece of the sacrum and coincides with the end of the sigmoid mesocolon. Having lost its mesentery—the mesorectum—during development, the rectum is covered by peritoneum only in front and along its sides. In the male, the peritoneum coming off the lower one-third of the rectum reflects onto the bladder to form the shallow **recto-vesical pouch**. In the female, parietal peritoneum passes first onto and over the uterus; in so doing, it creates the **recto-uterine pouch** (of Douglas) and the **broad ligament**. Douglas' pouch is important clinically, because it is the floor of the peritoneal cavity in the upright posture. Extravasated blood or other fluids may accumulate in it and become palpable on rectal examination.

The arterial supply to the rectum is via the **superior rectal artery**, the prolongation of the inferior mesenteric artery. The superior rectal artery divides opposite S3 into right and left branches which enter the rectum, subdivide, and supply the mucous membrane of the rectum and upper anal canal. Venous drainage is by way of the **superior rectal veins** whose continuation is the inferior mesenteric vein. More about the venous circulation of the anorectal segment in a later chapter.

The nerve supply to the rectum, like that of the rest of the hindgut, is from the inferior mesenteric plexus. It includes branches of the **pelvic splanchnic nerves**. Also known as *nervi eregenti*, they are preganglionic parasympathetic fibers from spinal cord segments S2–4.

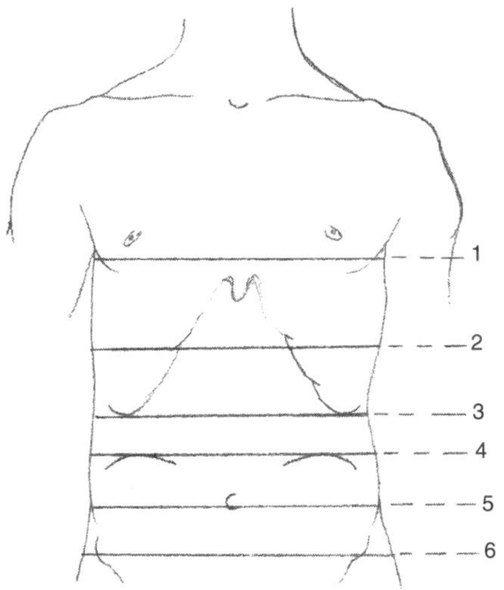

1. Xiphisternal plane (T8)
2. Transpyloric plane (L1)
3. Subcostal plane (L3)
4. Supracristal plane (L4)
5. Intertubercular plane (L5)
6. Interspinous plane (S2)

25.1 Transabdominal planes

1. Right and left kidneys
2. Pylorus and superior part of duodenum
3. Pancreas
4. Spleen
5. Fundus of gall bladder
6. Inferior vena cava
7. Superior mesenteric v.
8. Right and left colic flexures

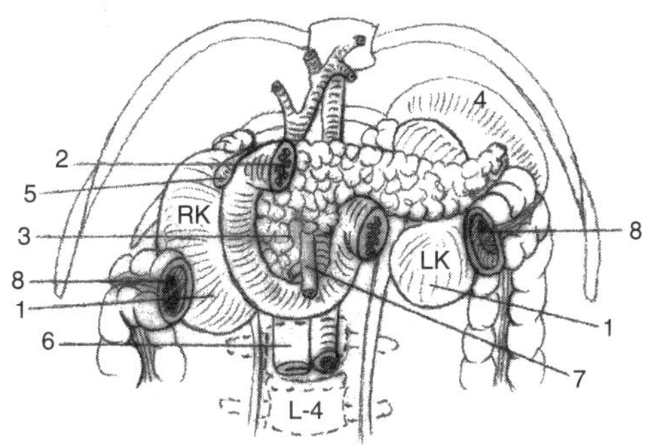

25.2 Topographic relation at transpyloric plane

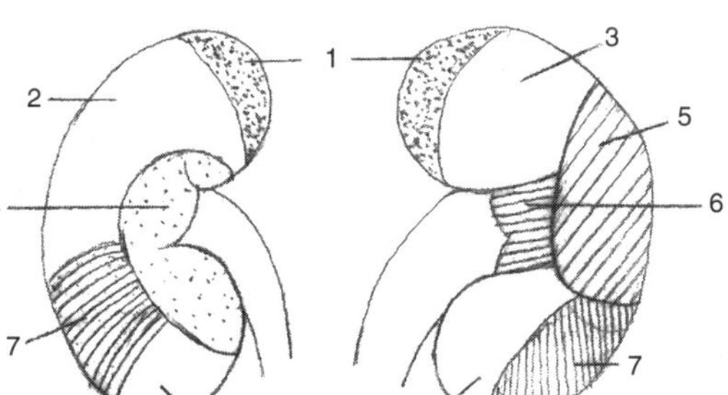

1. Adrenal
2. Liver
3. Stomach
4. Duodenum
5. Spleen
6. Pancreas
7. Colic flexure
8. Intestine

25.3 Anterior relations of kidney

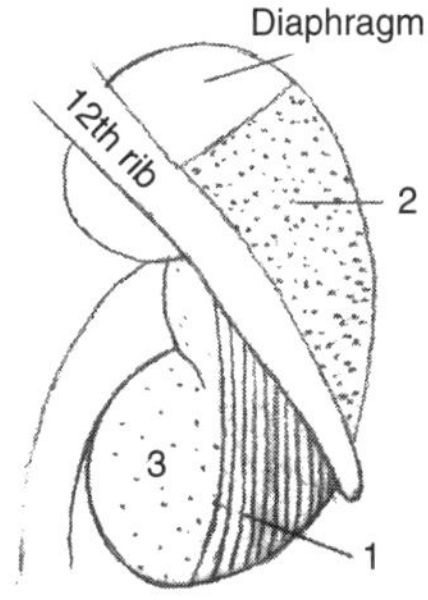

1. Quadratus lumborum m.
2. Transversus abdominis m.
3. Psoas major m.

25.4 Posterior relations of kidney

Figure 25.1: Transabdominal planes
Figure 25.2: Topographic relation at transpyloric plane
Figure 25.3: Anterior relations of kidneys
Figure 25.4: Posterior relations of kidney

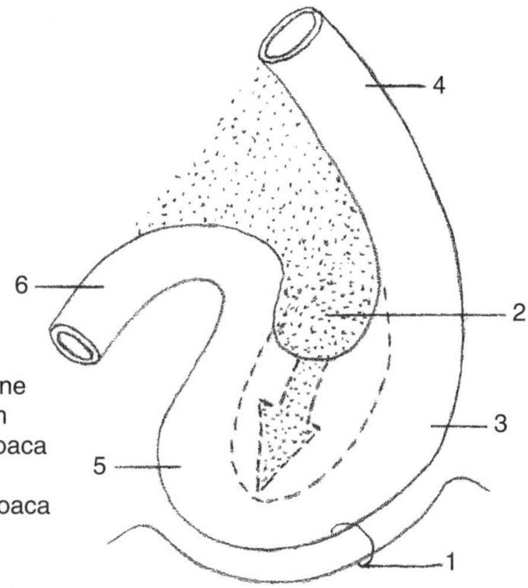

1. Cloacal membrane
2. Urorectal septum
3. Dorsal part of cloaca
4. Sigmoid colon
5. Ventral part of cloaca
6. Allantois

25.5 Cloaca and urorectal septum

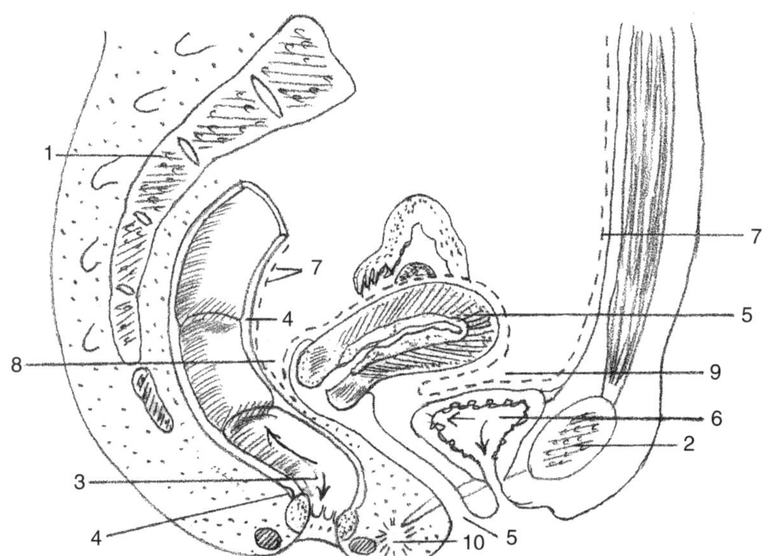

1. Sacrum and coccyx
2. Pubis
3. Rectum and anal canal
4. Sacral and anorectal flexures
5. Uterus and vagina

6. Bladder and urethra
7. Parietal peritoneum (dashed line)
8. Recto-uterine pouch (of Douglas)
9. Vesico-uterine pouch
10. Perineal body

25.6 Female pelvis (median section)

Figure 25.5: Cloaca and urorectal septum
Figure 25.6: Female pelvis (median section)

Chapter 26
FORMATION OF UROGENITAL SYSTEM

The primary source of the urogenital system is intermediate mesoderm. It first appears, you will recall, in the presomite embryo, during the third week of human development (Fig. 4.2, p. 25). Embryonic mesoderm simultaneously gives rise also to paraxial and lateral plate mesoderm. In the early embryo, intermediate mesoderm is located, bilaterally, between the other two forms of embryonic mesoderm; and, like paraxial mesoderm, it becomes transversely segmented without, however, differentiating into somite-like blocks of tissue.

Early Development

The embryo's first two attempts to develop excretory organs end in failure. Pairs of primitive kidneys, the **pronephroi** (Gr., plural of *pronephros*) form in intermediate mesoderm of the neck region during the fourth week of embryonic life (Fig. 26.1). They are composed of cords of cells, one pair per body segment, that link up at their tips laterally to form the **pronephric ducts**. Growing distally, the ducts contact the cloaca and open into its ventral portion one week later. The pronephroi soon degenerate, but the duct system remains intact.

As the pronephroi regress, the embryo develops a second pair of excretory organs, the **mesonephroi**. They also arise segmentally in intermediate mesoderm, but well below the neck region. The unit structures of the mesonephroi are the **mesonephric tubules**; they, too, link up distally with the pronephric duct, which is now renamed **mesonephric duct**. The mesonephroi function as primitive kidneys until the end of the second month. They also disappear eventually, leaving behind only a few mesonephric tubules and the pair of mesonephric ducts.

Mesonephric Ducts and Tubules

Both mesonephric tubules and ducts contribute significantly to the development of the *male* reproductive tract. The mesonephric tubules opposite the developing gonad (Fig. 26.2 a) establish connections with the future testis and persist as **efferent ductules**. The mesonephric ducts, also known as **Wolffian ducts**, are maintained between the levels of future testes and cloaca. They become the body's sperm-delivery system—*i.e.*, from proximal to distal, **epididymis, ductus** (or vas) **deferens**, and **ejaculatory duct. Seminal vesicles** form as outgrowths of the duct system.

In the *female* embryo (Fig 26.2 b), only the cloacal end of the mesonephric duct remains functional, for reasons that will soon become obvious. The rest of the system eventually atrophies; the tubules of epoopheron and Gartner's ducts are some of its inconsequential vestiges.

The Urogenital Ridges

In embryos five weeks old, the intermediate mesoderm bulges into the celom, the future peritoneal cavity, producing urogenital ridges (Fig. 26.3). Several important organs develop in the substance of the urogenital ridges of both sexes; among them, the permanent **kidneys**, the **gonads**, and the **adrenal glands**.

The formation of permanent human kidneys, the **metaneproi**, is the embryo's third attempt to develop a pair of excretory organs. Each metanephros arises from two distinctly separate sources: (1) the intermediate mesoderm of the urogenital ridges in the sacral region, the **metanephric blastema**; and, (2) the **ureteric** (nephrogenic) **bud**. The ureteric bud grows out of the cloacal end of the mesonephric duct—the part of the duct retained even in female embryos—and penetrates the metanephric blastema. Its blindly-ending tip undergoes repeated branching and induces the blastema to form a million or more nephrons, the structural and functional units of the kidney (Fig. 26.4). Without the organizing influence of the ureteric bud, nephrons will not develop. Beginning with the 10th week of intra-uterine life, the nephrons produce dilute urine. It is drained via collecting tubules—the microscopic, terminal divisions of the ureteric bud—into ducts of increasingly higher order. The stem of the original outgrowth persists as the **ureter**, and the first two divisions as **major** and **minor calyces**. All that remains to be said about the definitive kidney is that it "asends," relatively speaking, from the sacral location to a midlumbar position during the latter part of the embryonic period, changing its blood supply along the way. Each kidney also rotates in a manner such that its hilum eventually faces medially instead of ventrally.

Alongside the primitive digestive tube, the intermediate mesoderm of the urogenital ridge differentiates into the medullary (central) portion of the gonad (Fig. 26.3). The overlying celomic lining cell layer, *i.e.*, future mesothelium, forms the gonadal cortex, including the germinal epithelium. Remember that the celomic lining cell layer is derived from lateral plate mesoderm; hence, only the central parts of testis and ovary develop from intermediate mesoderm.

Celomic lining cells also form the cortex of the adrenal gland. The adrenal medulla, you will recall, develops from neural crest tissue which invades the dorsomedial aspect of the cortex towards the end of the embryonic period. The adrenal gland, therefore, has both mesodermal and ectodermal precursors.

The Paramesonephric Ducts

The paired **paramesonephric** (Muellarian) **ducts** arise towards the middle of the sixth week, each as a shallow cup on the lateral surface of the urogenital ridge (Fig. 26.3). The ducts grow tailwards in the substance of the urogenital ridge, lateral and parallel to the mesonephric ducts (Fig. 26.2 b). They cross in front of the mesonephric ducts, meet in the midline, and fuse dorsal to the ventral cloaca to form the **genital cord**—a vertical, solid column of cells (Fig. 26.5). Its blind tip ends by pushing up against the back wall of the ventral cloaca. The resulting slight bulge projecting into the interior of the ventral cloaca is **Mueller's tubercle**.

In *male* embryos, the paramesonephric ducts disappear except for upper and lower remnants—the appendix testis and prostatic utricle, respectively; neither one is noteworthy. The *female* embryo, however, claims the paramesonephric ducts for the development of its reproductive apparatus. Their unfused, vertical parts become **uterine tubes**, and the fused, horizontal segment of the duct system forms both fundus and body of the **uterus**. The genital cord develops into the uterine **cervix** and the upper portion of the **vagina**. The original funnel-shaped sites on the urogenital ridges that marked the beginnings of the paramesonephric ducts remain as the **ostia** (L., plural of *ostium* = mouth) of the uterine tubes through which the fimbriated end of each tube communicates with the peritoneal cavity.

You are correct to conclude, of course, that kidneys and ureters, adrenal glands, and internal reproductive organs all maintain their retroperitoneal positions when development is over.

The Cloaca

The **cloaca** is the expanded, terminal segment of the primitive digestive tube; hence, its germ layer source is embryonic endoderm. It is closed off below by the double-layered **cloacal membrane**, composed of cloacal endoderm internally and an external layer of cells which, like that of the orophayngeal membrane at the headend, is derived from ectoderm. The membrane separates the cloaca from the amniotic cavity. I have already alluded to the division of the cloaca into ventral and dorsal parts by downgrowth of the urorectal septum (Fig. 25.5, p. 178). Upon reaching the cloacal membrane towards the end of the sixth week of embryonic development, the urorectal septum fuses with the cloacal membrane and divides it into **urogenital** and **anal membranes** (Fig. 26.6). The anal membrane forms the floor of the dorsal cloaca, and the roof of the anal pit or proctodeum. When the anal membrane breaks down in the seventh week of gestation, dorsal cloaca (endoderm) and proctodeum (ectoderm) are confluent.

The dorsal cloaca, you will recall, becomes the epithelial lining of rectum and upper anal canal; further development of the ventral cloaca, however, is more complicated.

As shown in Fig. 26.6, the ventral cloaca is divisible into three compartments: the **urachus,** the **vesico-urethral pouch**, and the proper **urogenital sinus** (some textbooks of human embryology consider the vesico-urethral pouch to be the upper part of the primitive urogenital sinus).

The **urachus**, the uppermost chamber of the ventral cloaca, is the intra-abdominal continuation of the allantois. An early outgrowth from the primitive hindgut into the body stalk (Fig. 23.1, p. 164), the allantois later is incorporated into the umbilical cord. The urachus, hollow at first, becomes ligamentous when the cord is tied at birth; it is represented postnatally by the **median umbilical ligament** in the infra-umbilical abdominal wall.

The intermediate section of the ventral cloaca, the vesico-urethral pouch, is fated to become the epithelium of the **bladder**, the **bladder neck**, and the **primitive urethra**. In the male, the primitive urethra forms most of the prostatic urethra, and in the adult female accounts for about one-half of the organ's full length.

The epithelium of the rest of the male urethra is derived mainly from the urogenital sinus. An exception is that portion of the urethra contained in the glans penis; its epithelial germ layer source is not endoderm but ectoderm (see chapter 27). Sinus endoderm also forms the lining of the distal one-half of the female urethra and the epithelium of the lower vagina and vestibule.

Endodermal outgrowths from both vesico-urethral pouch and urogenital sinus develop into the parenchyma of glands associated with the reproductive system; among them, the **prostate** and **bulbo-urethral** (Cowper's) **glands** in the male, and the **urethral** and **vestibular** (Bartholin's) **glands** in the female.

Absorption of the Mesonephric Ducts

Let us now revisit the mesonephric ducts that terminate in the dorsolateral wall of the ventral cloaca of late-somite-stage embryos (Fig. 26.2). The short segment between the duct's cloacal end and the ureteric outgrowth is the **common excretory duct**, the final common path of the sperm-transport and urine-conveyal systems in the male. An imaginary line connecting the internal cloacal openings of the ducts defines the junction of vesico-urethral pouch and urogenital sinus. In frontal view (Fig. 26.7 a), the ventral cloaca's dorsal wall reveals single apertures; however, as each common excretory duct is gradually absorbed into the dorsal wall, mesonephric ducts and ureters ultimately will enter separately (Fig. 26.7 b). There follows rotation and shifting of positions, following which ureters open into the developing bladder, and (male) ejaculatory ducts into the future prostatic urethra (Fig. 26.7 c).

I will remind you that the epithelium of the cloaca is endodermal, whereas that of the mesonephric ducts and budding ureters is mesodermal. Absorption of the common excretory ducts, therefore, causes mesoderm to replace endoderm on the dorsal aspect of the vesico-urethral pouch. The area corresponds roughly to a triangle whose sides connect ureteric openings and entrance into the primitive urethra. The mesodermal contribution to the vesico-urethral pouch is said to be represented in its final form by the **trigone of the bladder**.

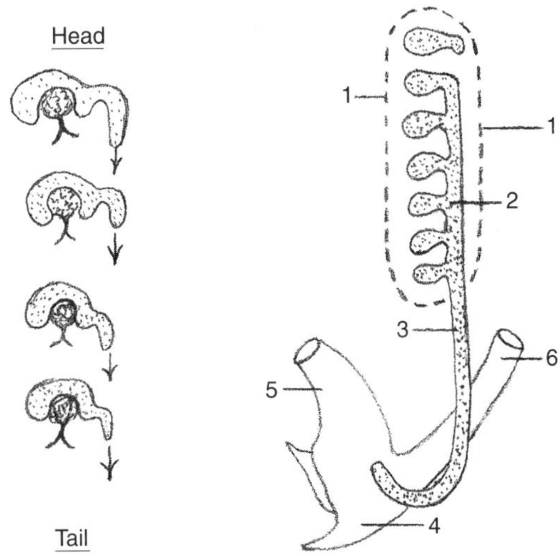

1. Degenerating pronephros
2. Pronephric cord
3. Pronephric duct
4. Cloaca
5. Allantois
6. Hindgut

26.1 Formation and fate of pronephros

Figure 26.1: Formation and fate of the pronephros

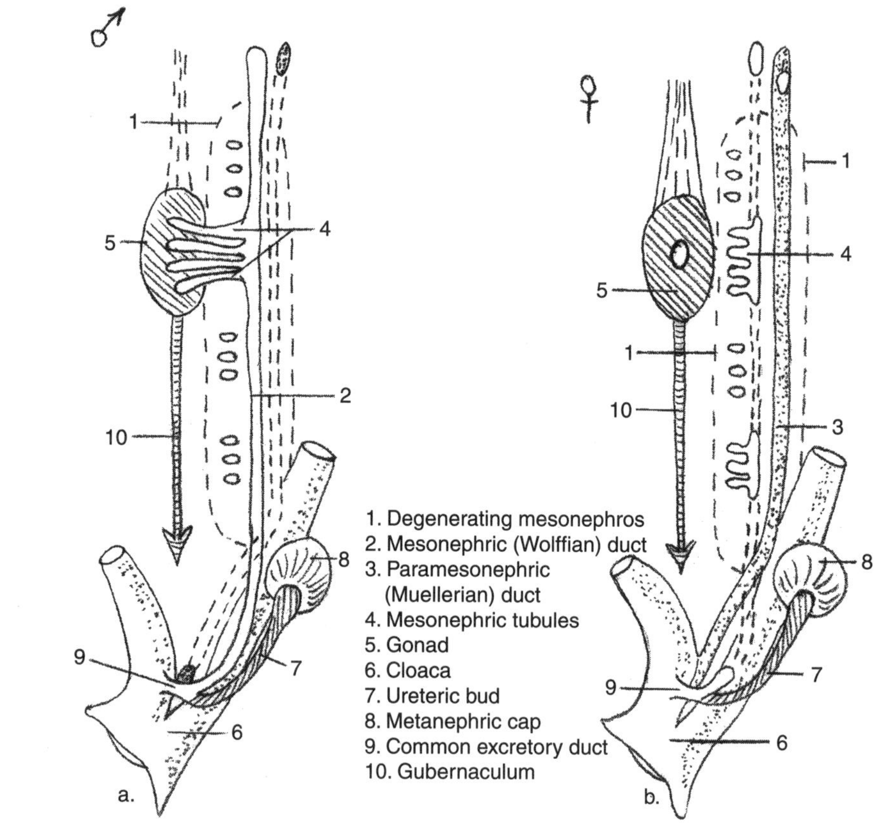

26.2 a-b Development of male and female duct systems

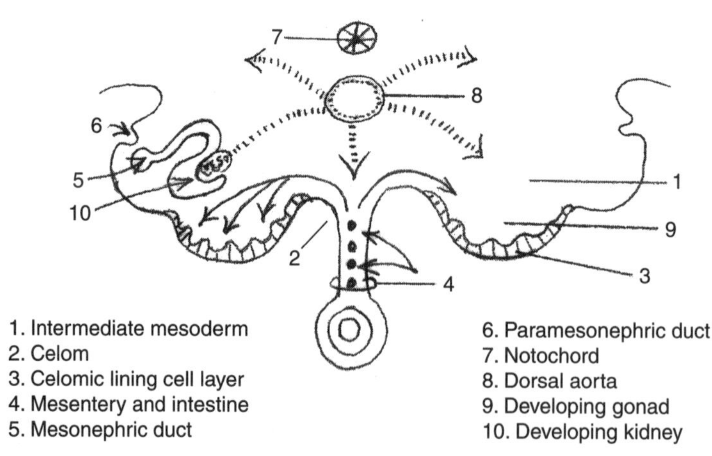

26.3 Urogenital ridges (cross-section)

Figure 26.2 a-b: Development of male and female duct systems
Figure 26.3: The urogenital ridges in cross-section

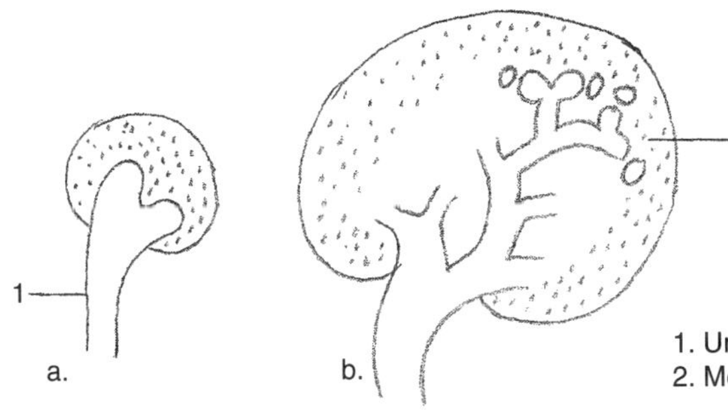

26.4 a-b Development of metanephros
a. Early stage b. Late stage

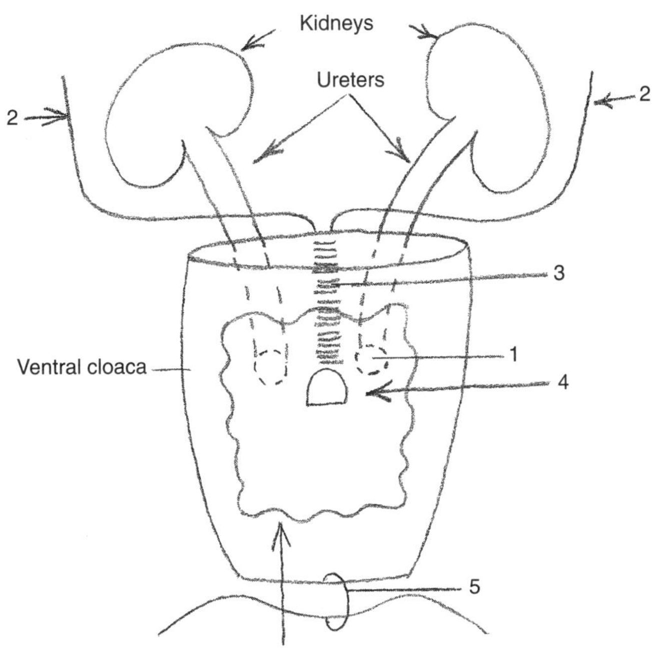

1. Entrance of ureter
2. Paramesonephric ducts
3. Genital cord
4. Mueller's tubercle
5. Urogenital membrane
6. Window in ventral wall

26.5 Dorsal wall of ventral cloaca (frontal view)

Figure 26.4 a-b: Development of metanephros
Figure 26.5: Dorsal wall of ventral cloaca in frontal view

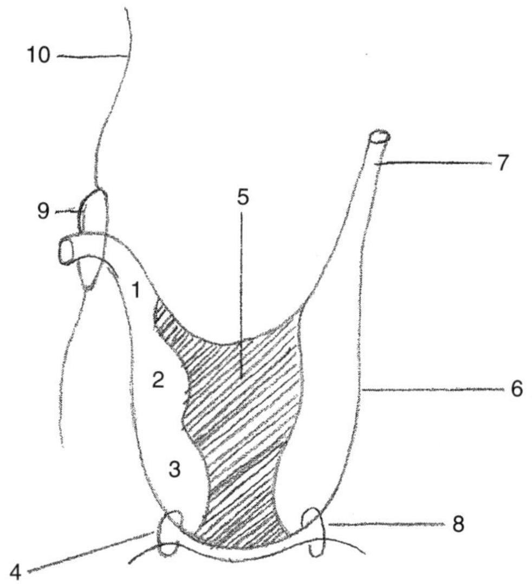

1. Urachus
2. Vesico-urethral pouch
3. Urogenital sinus
4. Urogenital membrane
5. Urorectal septum
6. Dorsal cloaca
7. Sigmoid colon
8. Anal membrane
9. Umbilical ring
10. Anterior abdominal wall

26.6 Divisions of ventral cloaca

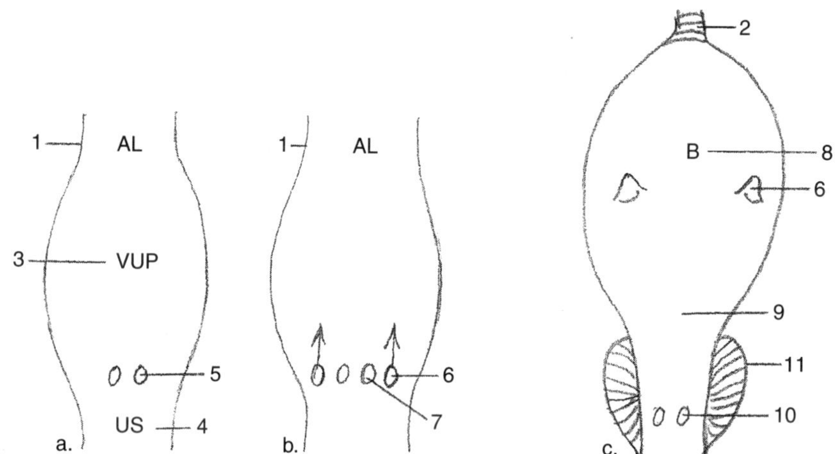

1. Allantois
2. Urachus
3. Vesico-urethral pouch
4. Urogenital sinus
5. Common excretory duct
6. Ureter
7. Mesonephric duct
8. Bladder
9. Primitive urethra
10 Ejaculatory duct
11. Prostate gland

26.7 a-c Stages in absorption of common excretory duct
a. Early b. Intermediate c. Late

Figure 26.6: Divisions of ventral cloaca
Figure 26.7 a-c: Absorption of common excretory duct

Chapter 27
MYOFASCIAL ANATOMY OF THE PERINEUM

The perineum is the subpelvic region, thus the "bottom" of the trunk. Its boundaries are: the pubic symphysis in front and the coccyx behind; the ischiopubic rami anterolaterally; and the paired sacrotuberous ligaments (between sacrum and ischial tuberosities) posterolaterally (Fig. 27.1).

The perineum contains the lower urethra, anal canal and sphincters, the roots of the external genitalia, and their support structures. It is separated from pelvic contents by the pelvic diaphragm which , therefore, forms both the floor of the abdominopelvic cavity and the roof of the perineum.

The Pelvic Diaphragm

The **pelvic diaphragm** is a sheet of striated muscle sandwiched between fascia. Muscle fibers and associated fascia of the pelvic diaphragm, like those of the abdominal diaphragm, have a circumferential bony origin and a central tendinous insertion. They arise peripherally from bones of the lesser (true) pelvis, then pass inferomedially, and converge upon the "central tendon" of the perineum, the **perineal body**.

The perineal body, a fibromuscular knot, is vital to the integrity of the pelvic diaphragm, and damage to it, if not repaired, will cause weakness of the pelvic floor and loss of structural support for pelvic viscera. The perineal body is important especially in the human female; for, being located between vagina and anal canal, it is liable to be torn during childbirth. More serious complication arise, of course, when the tear extends into the anorectal wall behind the perineal body. To forestall this from happening, many obstetricians favor a procedure, known as *episiotomy*, in which a small and controlled incision is made during labor and delivery in the posterolateral wall of the lower vagina. Such surgical intervention, it is claimed, not only shortens the period of labor and reduces the chance of fetal cerebral injury, but also simplifies wound closure and speeds healing. Incidentally, in clinical medicine, it customary to refer to the perineal body as the "perineum."

The pelvic diaphragm is composed of paired **levator ani** and **ischiococcygeus** (coccygeus, for short) **muscles**. Each levator ani muscle originates from the sidewall of the lesser pelvis between pubic bone in front and ischial spine behind. The coccygeus muscle fills in the rest of the pelvic diaphragm between ischial spine and coccyx.

Levator ani muscles are divisible into pubococcygeus and iliococcygeus parts (Fig. 27.1). Each **pubococcygeus muscle** arises from the back of the body of the pubic bone. The more anterior fibers of the pair pass along the vagina or prostate and form the sphincter-like **pubovaginalis** and **puboprostaticus** (levator prostatae) **muscles**, respectively. They end in the perineal body. Notice that the medial edge of neither muscle reaches the midline, thus allowing room for the passage of male and female urethrae and of the vagina into the perineum. The space between the free medial borders is the **urogenital hiatus**. It, clearly, is a weak spot in the pelvic floor; we shall soon learn how it is secured.

The intermediate fibers of the two pubococcygeus muscles, designated **puborectalis**, encircle the upper part of the anal canal. They form a U-shaped anorectal sling—the probable cause of the anorectal flexure. Lastly, the posterior fibers of the pubococcygeus muscles meet in the midline between anal canal and coccyx; the resulting raphe is the **anococcygeal body**.

Iliococcygeus, the other division of levator ani, arises partly from the pubic bone in front as a continuation of pubococcygeus; partly from the ischial spine behind; and partly from a thickening of obturator internus fascia—the **tendinous arch** (Fig. 27.2). The muscle inserts inferomedially into the anal canal between puborectalis and the external anal sphincter and into the anococcygeal body. Fibers of iliococcygeus contribute to the perineal body in both sexes.

Obturator internus muscle, while not part of the pelvic diaphragm, is a key structure as well. It is a broad, flat muscle that clings to the sidewall of the lesser pelvis (Fig 27.3). Obturator internus is a lateral rotator of the femur—but, in the female, its more important function may well be the padding it provides for the fetus during birthing.

Obturator internus is divisible into pelvic and perineal portions based on its relationship to the pelvic diaphragm. The line of demarcation is the tendinous arch which, as just described, gives rise to iliococcygeus.

The Anal Triangle

An imaginary line connecting the ischial tuberosities passes through the perineal body and divides the rhomboid-shaped perineum into two triangles of approximately equal size (Fig. 27.1). The **urogenital triangle**, in front, is occupied by perineal parts of the urethra and by roots of external reproductive organs, their muscles, vessels, and nerves. The **anal triangle** behind it contains the anal canal and external anal sphincter plus neurovascular supply lines. The rest of this chapter is devoted to the myofascial anatomy of the anal triangle.

The composition of the anal triangle is best displayed in frontal section (Fig. 27.3). Filling the iliac fossae are iliopsoas muscles covered by densely fibrous iliopsoas fascia; obturator internus muscles and fascia line the lesser pelvis. The pelvic diaphragm, represented here by the sloping fibers of iliococcygeus, has fascia on its superior and inferior aspects. The superior fascia of the diaphragm is an extension of abdominal transversalis fascia; it is continuous medially with the fascia over the rectum and other pelvic organs.

The inferior fascia of the pelvic diaphragm blends laterally with the obturator internus fascia and medially with the anal fascia over the external anal sphincter. It forms the roof of the **ischio-anal fossae.** Obturator internus fascia, the lateral wall of each fossa, splits into two layers near its lower edge to create the **pudendal** (Alcock's) **canal**. The pudendal nerve and internal pudendal vessels are carried forward inside the pudendal canal for distribution to all perineal organs.

The ischio-anal fossae contain large amounts of fatty connective tissue, the **ischio-anal fat pads,** enabling rectum, vagina, and other pelvic organs to expand sideways.

The Anatomy of the Anal Canal

The anal canal is approximately 3.75 cm (1.25 in.) long. It begins at the level of the pelvic diaphragm, the anorectal junction, descends into the perineum, and ends as the anus, 3.75 cm in front and below the coccyx.

Developmentally, the anal canal is divisible into cloacal and proctodeal parts. The cloacal part, the upper two-thirds of the canal, is derived, like the rectum, from the cloaca. Its mucosal epithelium, therefore, is endodermal in origin and, like the rest of the gut, composed of simple columnar cells. In contrast, the lining of the proctodeal (anal pit) part, the lower one-third of the canal, is stratified squamous epithelium and, like epidermis, ectodermal. The wall of the cloacal part of the anal canal is distinguished by six or more vertical folds of mucous membrane that extend upward into the lower rectum (Fig. 27.4). They are the **anal columns** (of Morgagni). The mucosal folds carry tributaries of the superior rectal vein. Stasis in the vessels may cause them to become varicose, causing anal columns, to bulge into the anorectal lumen as hemorrhoids.

Transverse mucosal folds join the lower ends of the anal columns to form **anal valves** enclosing **anal sinuses.** The scalloped ring so produced is the **pectinate line**, quite possibly the original site of the anal membrane and, thus, the boundary between developmentally distinct cloacal and proctodeal parts of the anal canal. The pectinate line is always in the cloacal part of the anal canal.

Just below the pectinate line is the **anal pecten** or **transition zone**, and just below it is **Hilton's white line**—a relatively narrow, avascular band of tissue separating mucosa above from epidermis below. The white line, or the pectinate line above it, is a watershed zone marking a distinct break in the territorial distribution of arteries, veins, lymphatics and nerves. The arterial supply above the zone is via the superior rectal artery, that below it by the inferior rectal artery. Venous blood above the white line leaves through both superior and middle rectal veins, that below it goes by way of inferior rectal veins. Lymph from tissues above the watershed flows upwards towards abdominopelvic lymph nodes; that from below it drains towards inguinal nodes. And, finally, innervation of the cloacal part of the anal canal is through fibers of the autonomic nervous system, whereas that of the proctodeal part is through somatic nerves—*e.g.*, inferior anal branches of the pudendal nerve.

Anal Sphincters

The **internal anal sphincter** is simply a thickening of the intramural, circular smooth muscle in the upper two-thirds of the anal canal. Its nerve supply is from both sympathetic and parasympathetic fibers of the autonomic nervous system.

The **external anal sphincter** surrounds the lower two-thirds of the anal canal. It is arranged in three concentric circles: the deep part of the external anal sphincter overlaps the internal sphincter, and superficial and subcutaneous parts complete the arrangement. The external anal sphincter being composed of striated muscle, its fibers are supplied by branches of the sacral plexus.

Thus, sphincters encircle the lumen of the anal canal from top to bottom. The internal and external anal sphincters form upper and lower two-thirds of the collar, respectively, and they overlap in its intermediate one-third.

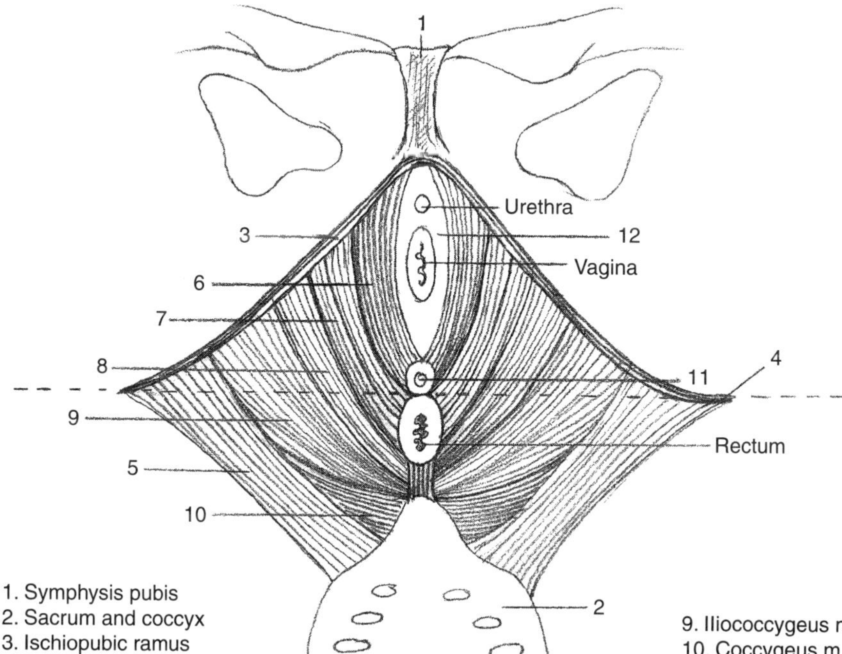

1. Symphysis pubis
2. Sacrum and coccyx
3. Ischiopubic ramus
4. Ischial tuberosity
5. Sacrotuberous ligament
6. Pubovaginalis (puboprostaticus) m.
7. Puborectalis m.
8. Pubococcygeus m.
9. Iliococcygeus m.
10. Coccygeus m.
11. Perineal body
12. Genital hiatus

27.1 Pelvic outlet and diaphragm

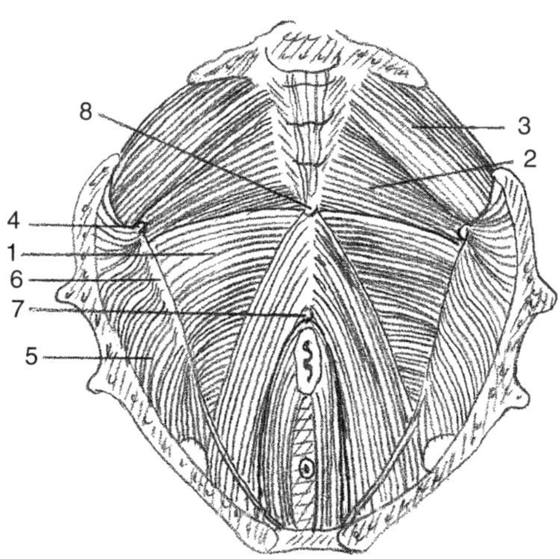

1. Levator ani m.
2. Coccygeus m.
3. Piriformis m.
4. Ischial spine
5. Obturator internus m.
6. Tendinous arch
7. Perineal body
8. Anococcygeal body

27.2 Interior of lesser pelvis (superior view)

Figure 27.1: The pelvic outlet and diaphragm
Figure 27.2: Interior of lesser pelvis viewed from above

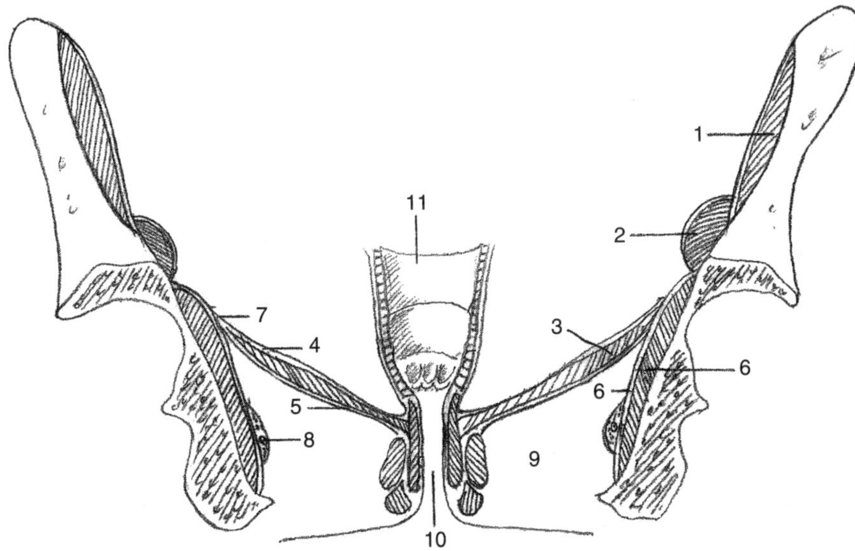

1. Iliacus m. and fascia
2. Psoas m. and fascia
3. Levator ani m.
4. Superior fascia of pelvic diaphragm
5. Inferior fascia of pelvic diaphragm
6. Obturator m. and fascia
7. Tendinous arch
8. Pudendal canal
9. Ischio-anal fossa
10. Anal canal and sphincter
11. Rectum

27.3 Anal triangle (frontal section)

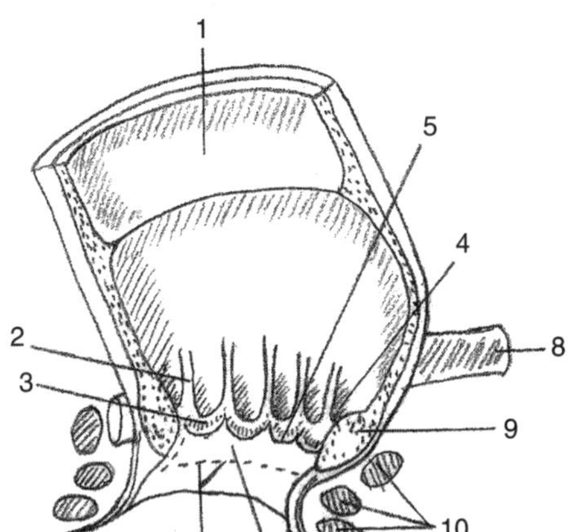

1. Rectum
2. Anal column
3. Anal valve
4. Anal sinus
5. Pectinate line
6. Anal pecten
7. Hilton's white line
8. Puborectalis m.
9. Internal anal sphincter
10. External anal sphincter

27.4 Anorectal segment (interior view)

Figure 27.3: The anal triangle in frontal section
Figure 27.4: Interior view of the anorectal segment

Chapter 28
THE UROGENITAL TRIANGLE

The **urogenital triangle** is another complicated and relatively inaccessible region of the human body; it is difficult to dissect and perhaps even more difficult to explain (now more so than ever, for the 1998 revision of the Nomina Anatomica has abandoned the term *urogenital diaphragm* to describe a key structure in the urogenital triangle). We'll start by taking a close look at the myofascial anatomy of the urogenital triangle, then review and amplify the developmental history of the urogenital sinus to explain sex differences in the region. The rest of the chapter will be devoted to a discussion of perineal spaces and fasciae relevant to clinical anatomy.

Myofascial Anatomy

Figs. 28.1 and 28.2 picture the myofascial anatomy of male and female urogenital triangles in the frontal plane. The section passes through the obturator foramina, closed in life by the obturator membrane, between superior and inferior pubic rami. Obturator externus and internus muscles overlie the membrane. Arising from the tendinous arch of obturator internus fascia, the levator ani muscles slant from lateral to medial, but never reach the midline. Observe, again, the urogenital hiatus between their medial free borders through which urethrae and vagina enter the perineum. In the male, the prostate gland and capsule fill the space. Note also that superior and inferior fasciae of the pelvic diaphragm are continuous through the hiatus.

As I pointed out previously, the urogenital hiatus is a weak spot in the pelvic floor. The risk of pelvic contents herniating into the perineum is greatly diminished, however, by the presence of a sheet of striated muscle and perineal membrane placed across and below the pelvic diaphragm. Muscle and membrane (until recently designated *urogenital diaphragm*) attach laterally to the ischiopubic rami as far back as the ischial tuberosities, but have free anterior and posterior borders.

The **perineal membrane** is composed of densely fibrous, white connective tissue, strong enough to anchor the root structures of penis and clitoris. Its anterior border, also known as the **transverse perineal ligament**, is separated from the inferior pubic ligament by a space that transmits the **deep dorsal vein** of the penis or clitoris, draining blood towards the pelvic venous plexus. The posterior border also is unattached except in the midline where it crosses the perineal body.

Fig. 28.3 is an inferior view of the muscle layer between ischiopubic rami. The **sphincter urethrae** is the external sphincter of the lower urinary tract and is under voluntary control. It merges above with the smooth muscle of the bladder neck—the involuntary, internal sphincter of the urinary bladder, innervated by autonomic nerves. With age, the fleshy fibers of the external urethral sphincter are replaced by connective tissue; the segment of urethra passing through the sphincter and perineal membrane, is the **intermediate** (membranous) **part** of the *male* urethra, intermediate between prostatic and spongy parts. It is the shortest and narrowest section of the organ. The **deep transverse perineal** muscles fill in the rest of the pubic arch.

In the *female*, whose pubic arch is wider than that of the male, the muscle and perineal membrane also are wider. In addition, the external urethral sphincter urethrae is modified so as to surround both vagina and urethra (sphincter urethrovaginalis). The deep transverse perineal muscle is represented only by smooth muscle.

The relatively inconspicuous superior fascia of the urogenital sphincter and deep transverse perineal musculature is an extension of obturator internus fascia (Fig. 28.1, 28.2). It blends medially with the inferior fascia of levator ani and forms the floor of the anterior recess of the ischo-anal fossa. The fascia, which adheres to the ischiopubic rami laterally, is continuous with the perineal membrane at the free anterior and posterior borders of the muscles.

Development of Perineal Organs

Fig. 28.4 is a ventral view of the tail end of a four-week-old human embryo. The cloacal membrane is still intact, but soon will be divided into urogenital and anal membranes by the descending urorectal septum. Fleshy folds of mesoderm and overlying ectoderm in the future perineal region fuse with the urorectal septum as it contacts the cloacal membrane. The **anal folds** at the margins of the anal membrane form, at first, the lateral walls of the anal pit and, later, of the lower anal canal and anus. The **genital folds**, flanking the urogenital membrane, join distally to form the **genital tubercle**, the primordium of the **glans penis** or **clitoris**. Anal and genital folds merge with the tip of the urorectal septum which persists as the perineal body.

Lateral to the genital folds is a second, more prominent, pair of skin thickenings. They are the **labioscrotal swellings**. Their fates differ in the two sexes: in the *male* embryo the swellings expand ventrally and unite to form the **scrotum**. They persist in the *female* as the paired **labia majora** (plural of *labium majus*).

As the genital tubercle lengthens to become the primitive **phallus**, the future penis or clitoris, genital folds and urogenital membrane also elongate (Fig. 28.5). Meanwhile, urogenital sinus endoderm has advanced into the genital tubercle, growing along its floor. And, when the urogenital membrane breaks down towards the end of the seventh week of human development, a dome-shaped plate of endoderm, the **urethral plate**, lies exposed to the exterior. It is continuous laterally with the ectoderm of the genital folds and skin surface. In the *male*, ventrally-directed growth of the genital folds brings about their fusion in the midline, thereby converting the urethral plate into a tube, the **spongy urethra**, while establishing the ventral (under) surface of the penis. The mucosal epithelium of the urethra contained in the glans penis, however, originates not from endoderm but ectoderm. It develops when cells on the phallus' surface multiply and grow into the glans (Fig. 28.6). At first a solid epithelial cord, the column of ectodermal cells subsequently canalizes, and its lumen links up with that of the spongy urethra.

In the *female*, the genital folds remain as the paired **labia minora** (plural of *labium minus*). They guard the entrance into the **vestibule**—the space into which urethra and vagina open. Developmentally, the vestibule and terminal urethra represent the lower portion of the urogenital sinus; both, therefore, are lined with endoderm. The lower vagina may be regarded as an outgrowth from the urogenital sinus (Fig. 28.7). It begins as an epithelial proliferation on the dorsal wall over Mueller's tubercle, the mound raised by the tip of the genital cord—the lower, vertically oriented, fused limbs of the paramesonephric ducts becoming uterine cervix. Condensation of sinus endoderm creates a solid **vaginal plate**, which invades the genital cord of mesoderm and moves it headwards. Towards midpregnancy, both vaginal plate and genital cord canalize; in this way, vagina and uterine cervix gain continuity. Evidently, then, the mucosal epithelium of the lower vagina is derived from urogenital sinus endoderm, whereas that of the upper vagina, like the epithelium lining the uterine cervix, has mesoderm as its source.

Both urethral and vaginal openings are brought near the body's surface when the urogenital sinus shortens and flattens to become vestibule.

The Clinical Anatomy of the Perineum

In the perineum, as elsewhere in the body, deep fascia is uniquely suited to delineate units of similar structure and/or common function. Four perineal fasciae combine to define three perineal compartments and contents in the urogenital triangle. The perineal spaces are important clinically, especially in the male, because extravasated blood or urine and pus may track along their fascial limits.

The **deep perineal space** or **pouch** (Figs. 28.1, 28.8) is bounded below by the perineal membrane. The pouch is occupied by afore-mentioned muscles, the intermediate urethra, and, in the male, by bulbo-urethral (Cowper's) glands and ducts. Vessels and nerves enter and leave it.

The **superficial perineal space** or **pouch** contains the root structures of penis or clitoris (Figs. 28.1, 28.2). The roof of the space is the densely fibrous perineal membrane, and its floor is the (deep) perineal fascia; more about it later. The **bulb of the penis**, and the paired **bulbs of the vestibule**, each covered with **bulbospongiosus muscle**, arise from the perineal membrane in the midline. Laterally, the membrane and ischial rami function to attach the **crura** of penis or clitoris and associated **ischiocavernosus muscles**. The crura, when traced forward, become **corpora cavernosa**—erectile tissue. The anterior continuation of the bulb in the male is the **corpus spongiosum**, bearing the spongy urethra and ending as **glans penis.** The **glans clitoridis** is formed by the two bulbs of the vestibule, which encircle the lower end of the vagina and unite in front of the urethral opening.

The superficial perineal pouch also contains vessels, nerves, and the **superficial transverse perineal muscles**. They, like the paired deep transverse perinei from which they are separated by the perineal membrane, help fix the perineal body. Curiously, the paired **greater vestibular** (Bartholin's) **glands**, the female developmental homologues of Cowper's glands, lie in the superficial, not the deep, perineal pouch.

The (deep) **perineal fascia** (Figs. 28.1, 28.8) forms the floor of the superficial perineal pouch; *i.e.*, it is *external* to the core structures of penis or clitoris. In the perineum, where it is also known as **Gallaudet's fascia**, the deep perineal fascia has the same shape and attachments as the perineal membrane with which it is continuous posteriorly around the superficial transverse perineal muscles. Anteriorly, the deep perineal fascia becomes the (deep) **fascia of the penis** (Buck's fascia), which fuses distally with the **tunica albuginea penis**, the sturdy membrane of white connective tissue that encloses the corpora cavernosa and spongiosum. The narrow cleft between Buck's fascia and tunica albuginea communicates proximally with the superficial perineal pouch.

The third perineal compartment, the **subcutaneous perineal pouch**, is a potential space, ordinarily empty. It separates the deep perineal fascia from the **subcutaneous perineal** (Colles') **fascia**. Like the other perineal fasciae, Colles' fascia attaches laterally to the pubic arch and fuses posteriorly with the free edge of the transverse perineal muscles—including the perineal body. In front, beyond the urogenital triangle, Colles' fascia becomes **tunica dartos** of the scrotal wall and septum, then the subcutaneous fascia of the penis. It continues into the abdomen as the membranous layer of the subcutaneous fascia of the lower abdominal wall, Scarpa's fascia, which fuses just below the inguinal ligament with the fascia lata of the thigh.

The subcutaneous perineal pouch, therefore, differs from both deep and superficial perineal pouches in being confluent in front with fascial spaces in the anterior abdominal wall and the subinguinal region. Traumatic injuries that puncture the deep perineal fascia and/or Buck's fascia, allow blood oozing from ruptured erectile tissue, or urine leaking from a torn urethra, to enter the subcutaneous perineal pouch. Blood or urine is then likely to spread to the scrotum, the penis, and into the anterior abdominal wall. In the absence of damage to the deep perineal fascia and/or Buck's fascia, however, extravasated fluids may diffuse only through the superficial perineal pouch and will be confined to it.

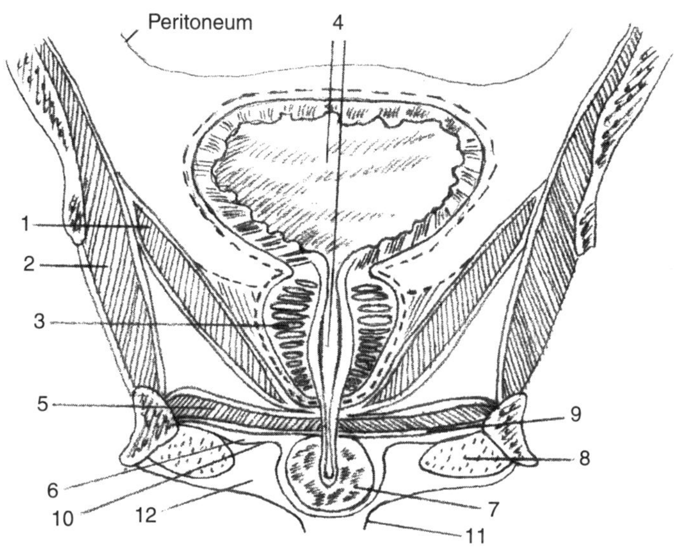

1. Levator ani m. and fascia
2. Obturator internus m. and fascia
3. Prostate gland in genital hiatus
4. Bladder and urethra
5. Deep perineal pouch and sphincter urethrae
6. Superficial perineal pouch
7. Bulb of penis with bulbospongiosus m.
8. Crus of penis with ischiocavernosus m.
9. Perineal membrane
10. (Deep) perineal fascia
11. Subcutaneous (Colles') perineal fascia
12. Subcutaneous perineal pouch

28.1 Male urogenital triangle (frontal section)

Figure 28.1: Male urogenital triangle (frontal section)

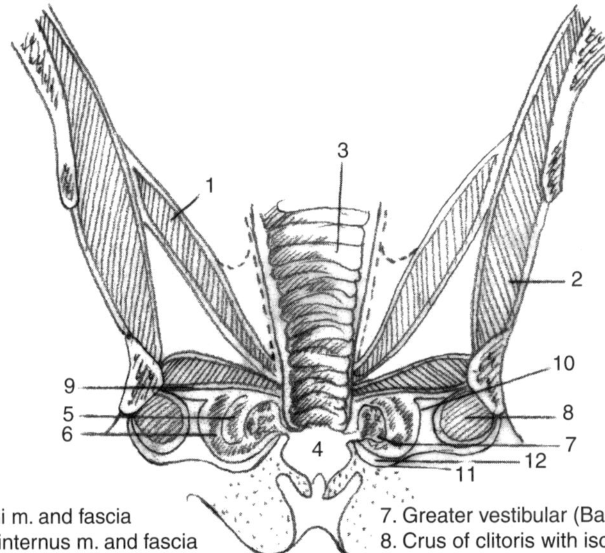

1. Levator ani m. and fascia
2. Obturator internus m. and fascia
3. Vagina in genital hiatus
4. Vestibule guarded by labia minora
5. Bulb of vestibule with bulbospongiosus m.
6. Superficial perineal pouch
7. Greater vestibular (Bartholin's) gland
8. Crus of clitoris with ischiocavernosus m.
9. Perineal membrane
10. (Deep) perineal fascia
11. Subcutaneous (Colles') perineal fascia
12. Subcutaneous perineal pouch

28.2 Female urogenital triangle (frontal section)

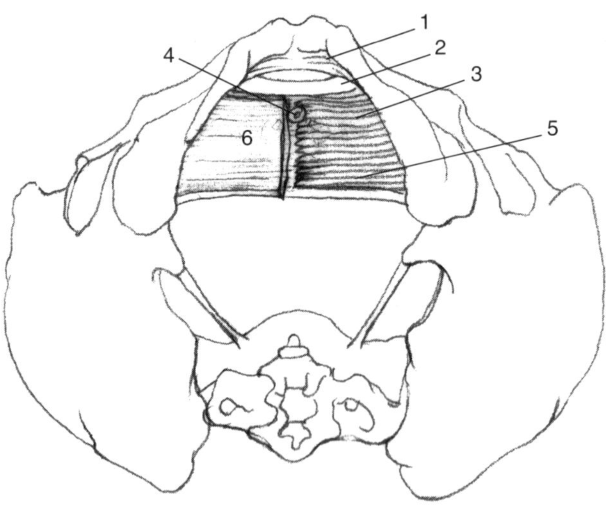

1. Inferior pubic ligament
2. Transverse perineal ligament
3. Sphincter urethrae (exposed)
4. Intermediate (membranous) urethra
5. Deep transverse perineal m.
6. Perineal membrane

28.3 Muscles of male deep perineal pouch

Figure 28.2: Female urogenital triangle (frontal section)
Figure 28.3: Muscles of the male deep perineal pouch

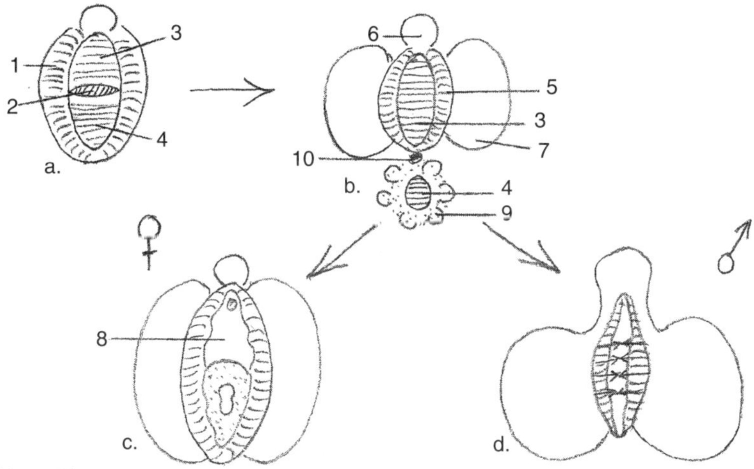

1. Cloacal fold
2. Urorectal septum
3. Urogenital membrane
4. Anal membrane
5. Genital folds
6. Genital tubercle
7. Labioscrotal swellings
8. Vagina and urethra in vestibule
9. Anal folds
10. Perineal body

28.4 a-d Developing perineum (ventral view)

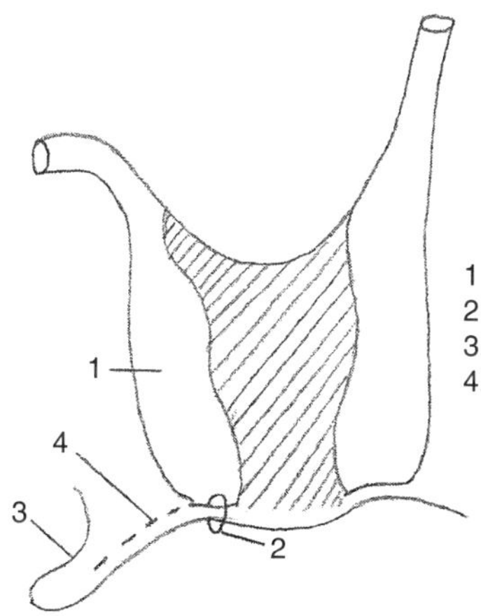

1. Urogenital sinus
2. Urogenital membrane
3. Phallus
4. Urethral plate

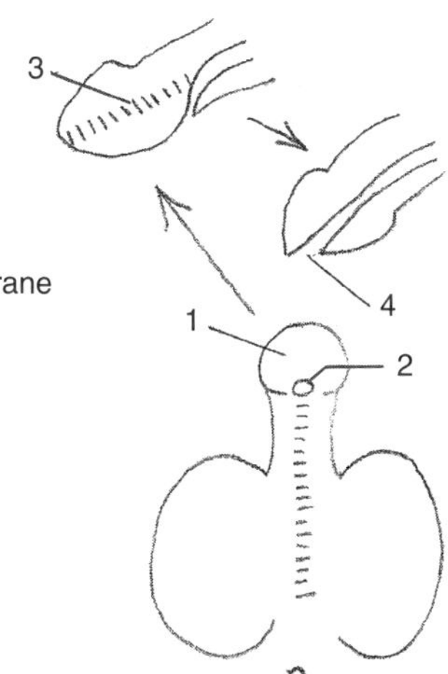

1. Glans penis
2. Temporary urethral opening
3. Epithelial cord
4. Permanent urethral opening

28.6 Formation of (male) glandular urethra

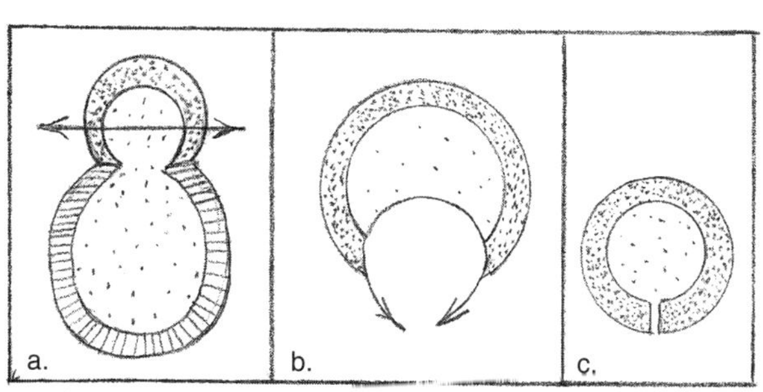

28.5 a-c Formation of (male) spongy urethra (median section)

Figure 28.4 a-d: Ventral view of developing perineum
Figure 28.5 a-c: Formation of (male) spongy urethra
Figure 28.6: Formation of (male) glandular urethra

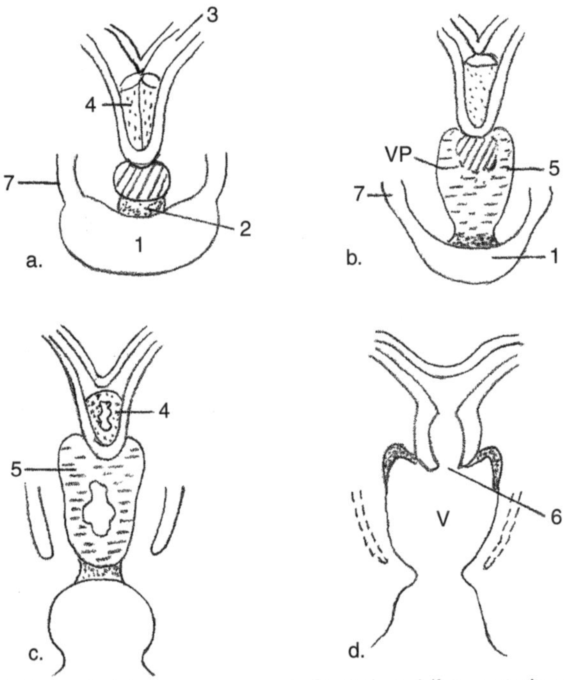

1. Urogenital sinus
2. Mueller's tubercle
3. Paramesonephric duct
4. Genital cord (future uterine cervix)
5. Vaginal plate
6. Uterovaginal passage after cord and plate canalize
7. Mesonephric duct

28.7 a-d Develoment of vagina

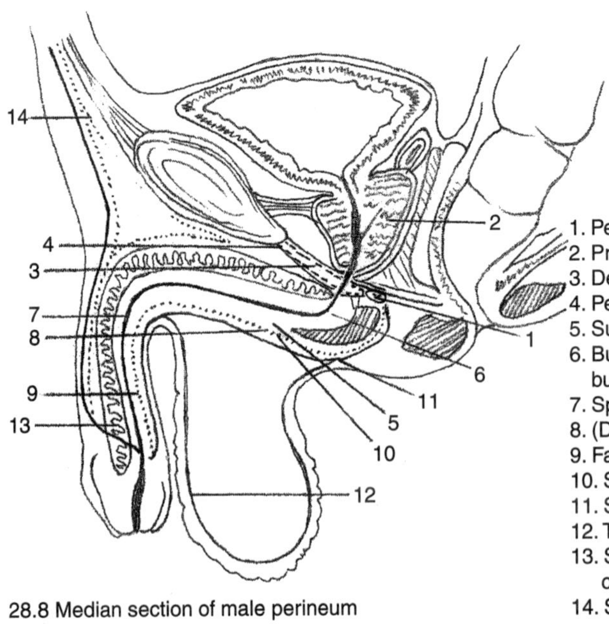

1. Perineal body
2. Prostate in genital hiatus
3. Deep perineal pouch and muscle
4. Perineal membrane
5. Superficial perineal pouch
6. Bulb of penis and bulbospongiosus m.
7. Spongy urethra
8. (Deep) perineal fascia
9. Fascia of the penis
10. Subcutaneous perineal pouch
11. Subcutaneous perineal fascia
12. Tunica dartos
13. Septum separating corpora cavernosa penis
14. Sarpa's fascia

28.8 Median section of male perineum

Figure 28.7 a-d: Stages in the development of the vagina
Figure 28.8: Median section of male perineum

Chapter 29
COLLATERAL CIRCULATION

You may take it as axiomatic that virtually all blood vessels form anastomoses to ensure the viability of the tissues of the body. Only so-called end arteries lack connections with other arteries, an accident of nature that poses serious threats to structures dependent for their survival on unimpeded flow through such vessel. In chapter 13, I referred at length to the collateral circulation of the head and neck, regions whose principal supply lines are the carotid and subclavian systems. We must now expand the discussion and consider the collateral circulation of other parts of the body.

Collateral Ciculation of the Upper Limb

The main arteries of the upper limb can be ligated at any point without blocking blood flow to distal parts of the extremity, because potential anastomoses exist between practically all branches of the subclavian-axillary system. The hub of the grid is the **scapular anastomosis**, shown in Fig. 29.1; only the vertebral branches of subclavian arteries fail to contribute vessels. On the other hand, the thoracic aorta communicates with both subclavian and axillary vessels through intercostal arteries, and the brachial artery connects with the scapular network through the deltoid branch of the profunda brachii. Even the carotid system, via the subclavian arteries, may channel blood into the upper limb by way of the scapular anastomosis.

Likewise, compression or occlusion of the brachial artery is no cause for real concern so long as blood flow through the profunda brachii artery remains unimpeded. And, if the brachial artery is ligated *above* the origin of the profunda brachii, the limb will continue to receive blood from the scapular anastomosis via the deltoid branch of the profunda brachii artery. Similarly, obstruction of the brachial artery at the elbow disrupts blood flow to forearm and hand only temporarily; collateral arteries arising proximal to the elbow joint anastomose with recurrent arteries originating distal to it and serve as alternative pathways (Fig. 29.2). Finally, a ligature placed around either of the forearm arteries is unlikely to have adverse long-term consequences, for both superficial and deep palmar arches are effective and efficient anastomoses.

Collateral Circulation of the Lower Limb

The principal lines of blood supply to the lower parts of the body, including limbs, operate through the iliac system. The common iliac arteries arise as terminal division of the abdominal aorta close to the supracristal (transumbilical) plane. They have no branches and, thus, resemble the common carotid arteries in the neck. Each common iliac artery divides into external and internal iliac arteries.

The **external iliac artery** supplies blood to the lower limb. It is renamed **femoral artery** as the vessel leaves the subinguinal region to enter the thigh at the midinguinal point (located on the inguinal ligament equidistant between anterior superior iliac spine and the pubic symphysis). In the upper thigh, the femoral artery gives off the large **profunda femoris branch**, then continues into the popliteal fossa, the region behind the knee joint, as **popliteal artery**. This part of the system ends by dividing into **anterior** and **posterior tibial arteries**.

The **internal iliac artery** has pelvic, perineal, gluteal, and femoral branches. Some branches, *e.g.*, the superior gluteal, inferior gluteal, and obturator arteries, anastomose with circumflex branches of the profunda femoris artery and, therefore, participate in the collateral circulation of the lower limb.

The basic plan of the system is build around four networks, shown schematically in Fig. 29.3. The most prominent of these, the **cruciate** (*i.e.*, cross-wise) **anastomosis**, is situated opposite the lesser trochanter of the femur. It receives input from the inferior gluteal artery above, from transverse branches of circumflex femoral arteries both medially and laterally, and from perforating branches of the profunda femoris below. The fourth, and last, perforating artery is part of the **knee joint anastomosis**. This vascular network is completed by superior and inferior genicular branches coming off the popliteal artery medially and laterally, by the descending branch of the lateral circumflex artery, and by the genicular branch of the femoral artery. A recurrent branch of the anterior tibial artery enters into the anastomosis from below.

Above the level of the cruciate anastomosis, the symmetry of vascular connections is lost. The ascending branch of the medial circumflex artery anastomoses with superior gluteal, inferior gluteal, and obturator arteries over the greater trochanter. That of the lateral circumflex artery links up with superior gluteal vessels and with branches of both deep and superficial circumflex iliac arteries near the anterior superior iliac spine.

The **anterior tibial artery**, becoming dorsalis pedis over the dorsum of the foot, and the lateral plantar branch of the **posterior tibial artery** anastomose at the bottom of the foot to form the **plantar arch**.

Collateral Circulation of the Body Wall

In chapter 22, I referred to the anastomosis in the anterior abdominal wall that shunts blood from the subclavian artery directly into the external iliac artery. It has its counterpart in the venous system, in which the inferior epigastric vein of the external iliac anastomoses with the superior epigastric vein of the internal thoracic inside the rectus sheath. This channel becomes operational when ligation of the inferior vena cava is indicated for the prevention of pulmonary embolism from venous blood clots (phlebothrombosis) that originate in the lower extremities.

A superficial anastomosis in the anterior body wall unique to the venous circulation, known as the **thoraco-epigastric system**, joins femoral and axillary veins (Fig. 29.4). It connects tributaries of the *superficial* epigastric vein, a vessel ending in the great saphenous vein, with those of the lateral thoracic vein. Tributaries of the internal thoracic may be part of the system and return blood to the right atrium via brachiocephalic veins.

On a deeper plane, venous blood may circumvent the inferior vena cava by a pathway uniting common iliac veins and the azygos system. Collateral circulation is realized through iliolumbar, ascending lumbar, and subcostal veins (Fig. 29.5). Be advised, however, that the ascending lumbar veins communicate by way of intervertebral foramina with spinal and, hence, vertebral veins. Thus, neoplastic disease, such as cancer of the prostate, having invaded the pelvic venous plexus and reached the ascending lumbar veins, may disseminate to the brain. Malignancy cells travelling this route also may end up in the lungs, once they enter the bronchial vessels of the azygos system.

Portocaval Anastomoses

These anastomoses act to shuttle blood between systemic and portal circulations. Thus, when venous blood returning from the pelvis and lower limb is blocked from entering the inferior vena cava, it may still reach the heart by way of the portal system. Conversely, in obstructions to blood flow through the portal vein or its tributaries, portal venous blood is forced into the systemic circuit. You should know, however, that portocaval anastomoses are, at best, sites of sluggish circulation. If exposed to retrograde overloads, they tend to sequester blood and produce clinical signs. The most important portocaval anastomoses are at the gastro-esophageal and anorectal junctions.

The **portal vein**, you will recall, transports venous blood from all parts of the digestive system to the liver. When passage of blood through the liver or portal vein is compromised, portal venous pressure rises. Conditions producing *portal hypertension* include fibrocytic degeneration or cirrhosis of the liver, compression of the superior mesenteric vein due to cancer of the pancreas, and enlargement of hepatic lymph nodes in the lesser omentum. Ascites, the accumulation of fluid in the peritoneal cavity attributable to leaking vessels, is symptomatic of portal hypertension.

The anastomosis in the **anorectal segment** links the superior rectal veins carrying blood to the portal system with middle and inferior rectal veins draining into the internal iliac (Fig. 29.5). The veins ascending in the anal columns, which terminate in the superior rectal veins, communicate with a venous plexus outside the segment through thin-walled vessels that pass through the anorectal wall at right angles to its long axis. Even under normal circumstances, blood flow through them is intermittent, and it comes to a halt when the circular smooth muscle of the anorectal wall contracts. Compression of the ascending veins during defecation or "bearing down," *etc.*, may produce engorgement and distension of the anal columns, causing them to protrude into the lumen of the anal canal as hemorrhoids or "piles." If grossly enlarged, hemorrhoids are apt to bleed when torn by a fecal mass on the way out. Obviously, the danger of such varicosities developing, and of severe hemorrhage occurring, is magnified many times over during portal vein obstruction, when retrograde flow and increased venous pressure exacerbate the problem.

At the **gastro-esophageal junction**, the portocaval plexus is formed by tributaries of esophageal veins draining into the azygos system (systemic circulation) and tributaries of gastric veins terminating in the portal vein (Fig. 29.5). Blood moves slowly through the venous plexus; the gastro-esophageal junction, therefore, is susceptible to the same complications that affect the lower end of the gut. Portal hypertension, when present, elevates folds of mucous membrane containing greatly distended vessels. They bulge into the lumen of the abdominal esophagus, where they may be shorn off as a bolus of food passes over them. The development of varicosities at the gastro-esophageal junction is a very serious matter; for, bleeding from "esophageal piles" is not easily controlled and, indeed, may be life-threatening.

A third portocaval anastomosis joins the portal vein and the thoraco-epigastric system of the abdominal wall through para-umbilical veins travelling with the round ligament between liver and umbilicus (Fig. 29.5). Portal hypertension drives blood into the anterior body wall, raising a web of tortuous vessels that radiate outwards from the umbilicus like spokes of a wheel. This pattern of venous markings on the body's surface is known as **caput Medusae**, the head of the Medusa, after the mythological figure whose scalp sprouted serpents instead of hair.

A portocaval anastomosis exists also in the **posterior body wall**. Here, retroperitoneal veins (of Retzius) collect blood from the bare area of the liver and deposit it in the inferior vena cava. They communicate with the portal vein through small vessels in the body wall draining blood from parts of the large intestine.

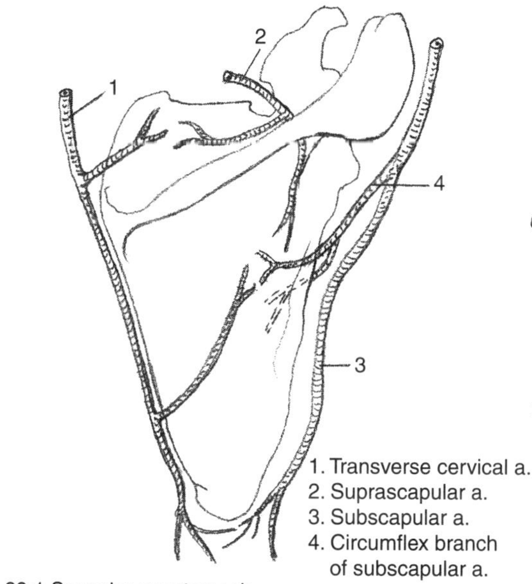

29.1 Scapular anastomosis

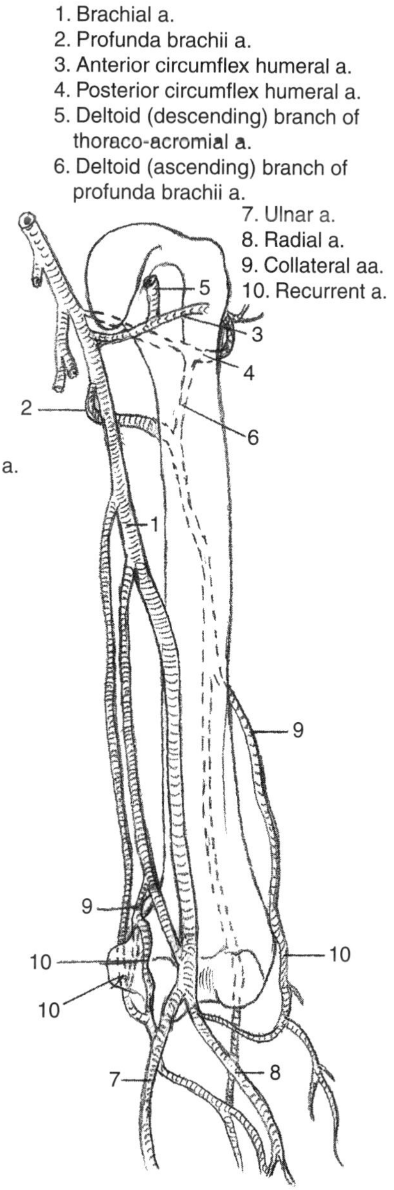

29.2 Elbow joint anastomosis

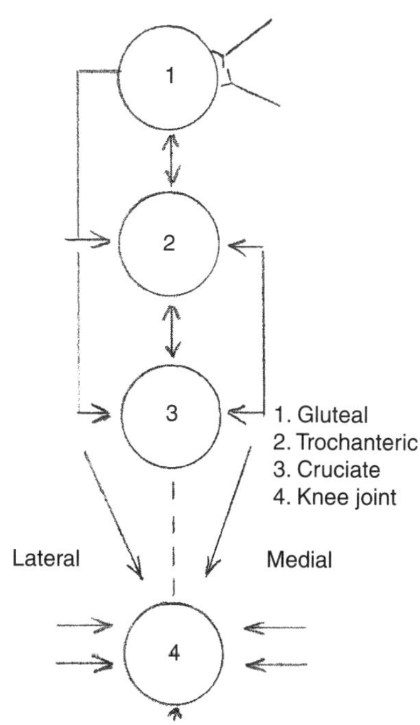

29.3 Anastomoses of lower extremity (see text for detail)

Figure 29.1: The scapular anastomosis
Figure 29.2: The elbow joint anastomosis
Figure 29.3: Anastomoses of the lower extremity

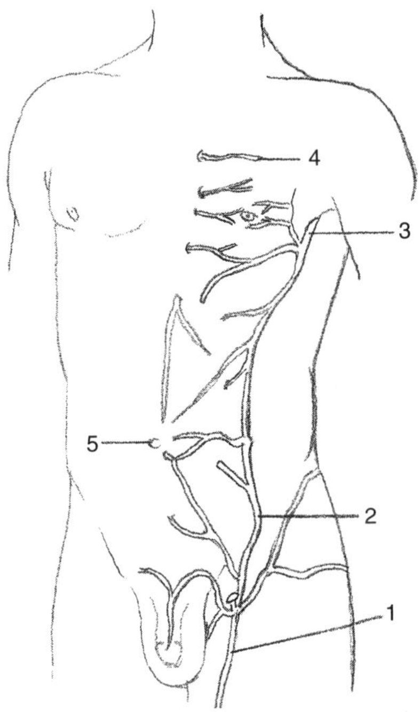

1. Great saphenous v.
2. Superficial epigastric v.
3. Lateral thoracic v.
4. Anterior intercostal v.
5. Venous convergence at umbilicus

29.4 Thoraco-epigastric system

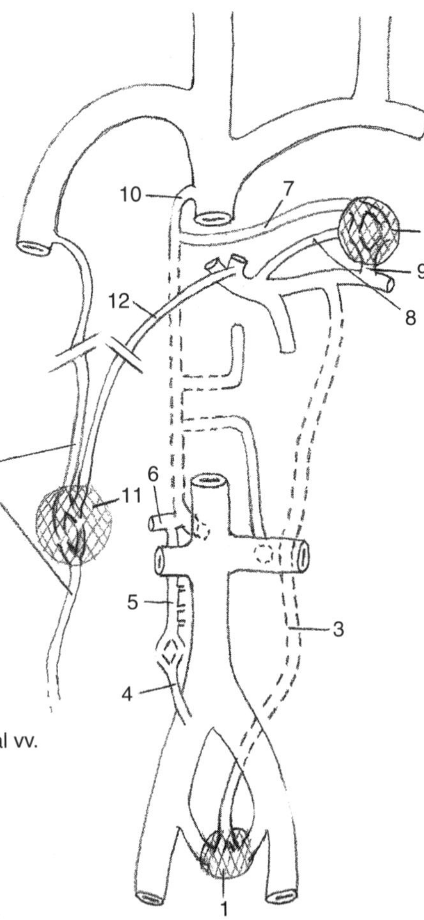

1. Anorectal junction and venous plexus
2. Gastro-esophageal junction and venous plexus
3. Inferior mesenteric v.
4. Iliolumbar v.
5. Ascending lumbar v.
6. Subcostal v.
7. Esophageal v.
8. Left gastric v.
9. Short gastric v.
10. Arch of azygos v.
11. Caput Medusae
12. Round ligament of liver and transumbilical vv.
13. Thoraco-epigastric v.

29.5 Portocaval anastomoses

Figure 29.4: The thoraco-epigastric system
Figure 29.5: Portocaval anastomoses

Chapter 30
FIELDS OF LYMPHATIC DRAINAGE

This section is an extension of chapter 17, which introduced you to the descriptive, functional, and clinical anatomy of the lymphatic system. We now continue that discussion by examining vessels and nodes engaged in the lymphatic drainage of the thorax, abdomen, and pelvis, and conclude with the general plan of lymphatics draining upper and lower limbs.

Lymphatic Drainage of Thorax

Lymph of the *left* hemi-thorax enters the **thoracic duct**, either directly or indirectly by way of left **bronchomediastinal** and **descending intercostal trunks.** Lymph of the *right* hemi-thorax takes a different path (see chapter 17). The nodes receiving lymph from the thoracic wall and upper intercostal spaces are related in front to the internal thoracic vessels, **parasternal nodes**; and in back, as shown in Fig. 30.1, to the thoracic aorta, **intercostal** and **prevertebral nodes**. **Superior** and **inferior diaphragmatic nodes** filter lymph from the lower intercostal spaces, the diaphragm, the bare area of the liver, and the gastro-esophageal junction. Lymph leaving these nodes enters the descending intercostal trunks which, after passing through the aortic hiatus, terminate in the abdominal thoracic duct or cisterna chyli. Note the presence of a subdiaphragmatic lymphatic plexus in the parietal peritoneum. Its vessels appear to be in free and open communication with the peritoneal cavity; intraperitoneal injections or infusions, therefore, are effective means to speed absorption of drugs, or even whole blood, into the body's circulation. Unfortunately, however, in cases of peritonitis due to bacterial invasion, contaminated peritoneal fluid may spread by this route to the lymphatic system and beyond.

Brachiocephalic, pericardial, and **juxta-esophageal nodes** receive lymph from thoracic viscera. Lymphatic drainage from the lungs is remarkable, in that lymph must negotiate a three-tier system of nodes before reaching the bronchomediastinal trunks. Lymph first encounters **intrapulmonary nodes** related to the divisions of the main bronchi. It next flows alongside pulmonary vessels to the **bronchopulmonary** (hilar) **nodes** at the root of the lungs, from which it is carried to the **tracheobronchial nodes** that surround the bifurcation of the trachea. Superficial lymph originating in the visceral pleura also drains centripetally towards the hilar nodes. The lymphatic vessels of this plexus communicate with the deep lymphatics of the lung, and lymph is diverted towards them when pulmonary disease disrupts internal lymphatic outflow.

Both bronchopulmonary and tracheobronchial lymph nodes usually contain carbon deposits, thus are radio-opaque and visually identifiable on X-ray examination.

Lymphatic Drainage of Abdomen

Groups of regional **nodes**, named **gastric, hepatic, pancreatic, splenic,** *etc.*, filter lymph from submucous and subserous tissues of the alimentary canal, and from the liver, gall bladder, pancreas, and spleen. Lymph from these outlying nodes, flows retroperitoneally towards groups of abdominal lymph nodes surrounding the roots and stems of the unpaired gut arteries—*i.e.*, celiac, superior and inferior mesenteric arteries (Figs. 30.1–30.3). These arteries, you will recall, have very specific fields of distribution, each vessel and its branches supplying organs developed from one of the early embryo's three gut segments. Fields of lymphatic drainage are based on similar developmental criteria. Thus, **celiac nodes** receive lymph from derivatives of the primitive foregut, including stomach, duodenum, pancreas, liver-gall bladder, and spleen. The **superior mesenteric nodes** filter lymph from organs that developed from the embryonic midgut. **Inferior mesenteric nodes** intercept lymph coming from all parts of the primitive hindgut region, including the cloacal part of the anal canal.

Lymph collected from the alimentary canal will have passed through a series of nodes before reaching a pre-aortic group. It first meets aggregated lymphoid nodules in the gut wall from pharynx to anus, next drains to **mesocolic nodes** located at the junction of mesentery and gut wall (Fig. 30.2). Lymph then advances towards **mesenteric nodes**, alongside jejuno-ileal and colic arteries. An exception is made for lymph from lacteals of intestinal villi. It, laden with emulsified fat after a meal, bypasses all lymph nodes and goes directly to the cisterna chyli via one or more **intestinal trunks**.

The rest of the abdominal lymph is brought eventually to the celiac nodes, from which efferent vessels deposit it into the abdominal part of the thoracic duct. Note, however, that efferents from a few inferior mesenteric nodes transport lymph directly into the left lumbar trunk.

Here as elsewhere, enlarged and/or infected lymph nodes may adversely affect neighboring structures. Carcinoma of the pylorus, for instance, frequently involves the retropyloric lymph nodes, causing lymph to back up into hepatic nodes which then become sites of malignant growth. In such cases, pressure on the biliary duct system may produce (obstructive) jaundice, and compression of the portal vein may result in portal hypertension.

Lymph from paired retroperitoneal organs, such as adrenal glands, kidneys, and ureters is conveyed to chain-linked **left lateral aortic** and **right lateral caval nodes** (Fig. 30.3). Their efferent vessels form the paired lumbar trunks that become thoracic duct. Below, the abdominal nodes are interconnected with pelvic lymph nodes that surround the common iliac arteries and their branches.

Lymphatic Drainage of Pelvis

Internal iliac nodes receive lymph from pelvic organs, and **external iliac nodes** are the destination for lymph from the lower extremities (Fig. 30.3). The two streams converge onto common iliac nodes, pass through the right and left lumbar nodes, then enter the lumbar trunks.

Lymph of rectum and anal canal *above* Hilton's white line (the mucocutaneous junction) is brought to the inferior mesenteric nodes by lymphatics that accompany the superior rectal arteries, but it just as frequently passes into the internal iliac nodes. Lymph collected from *below* Hilton's white line drains along the inferior rectal arteries to the superomedial group of **superficial inguinal nodes**. These nodes receive lymph also from perineal organs, such as the urethra, lower vagina, and external genitalia. Lymph from the glans penis and clitoris, however, goes uninteruptedly to deep inguinal nodes, which relay lymph to external iliac nodes. Ultimately, perineal lymph ends up in the thoracic duct. It gets there by the same circuitous route that is travelled by superficial lymph from the lower limb, as you shall see.

Lymphatics of the upper vagina, uterine cervix and of most of the rest of the uterus carry lymph to the internal iliac nodes. Some vessels, however, draining lymph from the upper body and fundus of the uterus, take the course of the round ligament and, thus, deliver lymph to the superficial inguinal nodes. Collecting vessels from the bladder, seminal vesicles, and prostate gland pass either to internal or to external iliac nodes.

Lymph flow from the testes and ovaries, including uterine tubes, follows the path of testicular and ovarian arteries; *i.e.*, lymph is brought to abdominal rather than to internal iliac or inguinal lymph nodes. Lateral aortic and caval nodes vary in number and location, but one right lateral node has a fairly constant position below the angle formed by the entrance of the right renal vein into the inferior vena cava. It is apt to become involved in metastatic cancer of the right testis and ovary.

The **middle sacral** (promontorial) **nodes**, clustered alongside the median sacral artery, and a group of nodes situated opposite vertebra S3, within the bifurcation of the superior rectal artery, also deserve special mention. Most ascending rectal lymph vessels pass in their direction, and all must be removed in operations for cancer of the rectum.

Lymphatic Drainage of Upper and Lower Limbs

Patterns of lymphatic drainage are similar in upper and lower limbs: superficial lymph vessels accompany superficial veins, and deep lymphatics travel with major arteries.

All of the lymph from the upper limb ends up in the **axillary nodes**, though it arrives there by different routes. Superficial lymph vessels run with cephalic and basilic veins placed, respectively, at the limb's lateral (radial or pre-axial) and medial (ulnar or postaxial) borders. Thus, lymphatics from the radial side of hand, forearm, and arm pass with the cephalic vein to the deltopectoral nodes (see chapters 7 and 17; also Fig. 7.4, p. 46); they, then, penetrate the clavipectoral fascia to reach the apical nodes of the axillary group. Superficial lymph from the ulnar side of the limb flows along the basilic vein and is intercepted by **supratrochlear lymph nodes** in the cubital fossa. It then tracks with the basilic vein as that vessel pierces the deep fascia to become axillary vein. Surface and deep tissue lymph join at this point, and the combined stream is brought to the lateral group of nodes on the axillary vein. From there, efferent vessels carry lymph to the apical nodes and into the subclavian lymph trunk.

Note that lymph from the radial side of the limb encounters few nodes during its ascent into the axilla compared to lymph from the ulnar side. This fact may explain why infections of the thumb spread more rapidly to the axillary nodes than do those of the little finger.

In the lower limb, most of the surface lymph is channeled alongside the great saphenous vein to the inferior group of superficial inguinal nodes strung out lateral to the vessel. Lymph then proceeds to the **deep inguinal nodes** (of Cloquet), located in the femoral canal under the fascia lata, the deep fascia of the thigh. As their afferent lymphatics pass through the saphenous opening in company with the saphenous vein (Fig. 30.3), the vessels perforate the overlying superficial fascia which, hence, acquires the name of cribriform fascia (L., *cribrum* = a sieve).

While most of the lymph from the surface of the limb drains in the direction taken by the great saphenous vein, superficial lymph from the area of the heel follows the route of the small saphenous vein to the popliteal fossa. Here, after penetrating the deep fascia, this lymph stream reaches the **popliteal nodes**. These same nodes also receive lymph from the deep tissues of the foot and calf. Effluent popliteal lymph, travelling the path of the femoral artery and vein, is transported to the deep inguinal nodes, where it is joined by lymph draining from the deep tissues of the thigh. The deep inguinal nodes, therefore, are the end station for all lymph from the lower limb, the perineum, and the superficial tissues of the lower back. The bulk of this lymph passes out of the pelvis and into the abdomen alongside the external iliac and common iliac vessels, then enters the lumbar trunks.

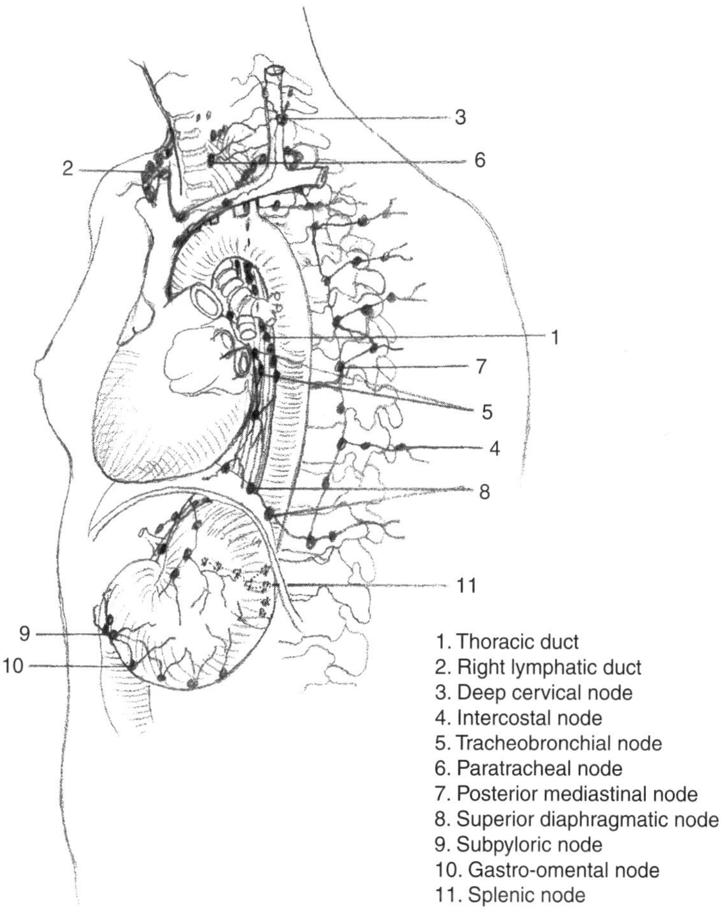

30.1 Lymphatic drainage of thorax

Figure 30.1: Lymphatic drainage of thorax

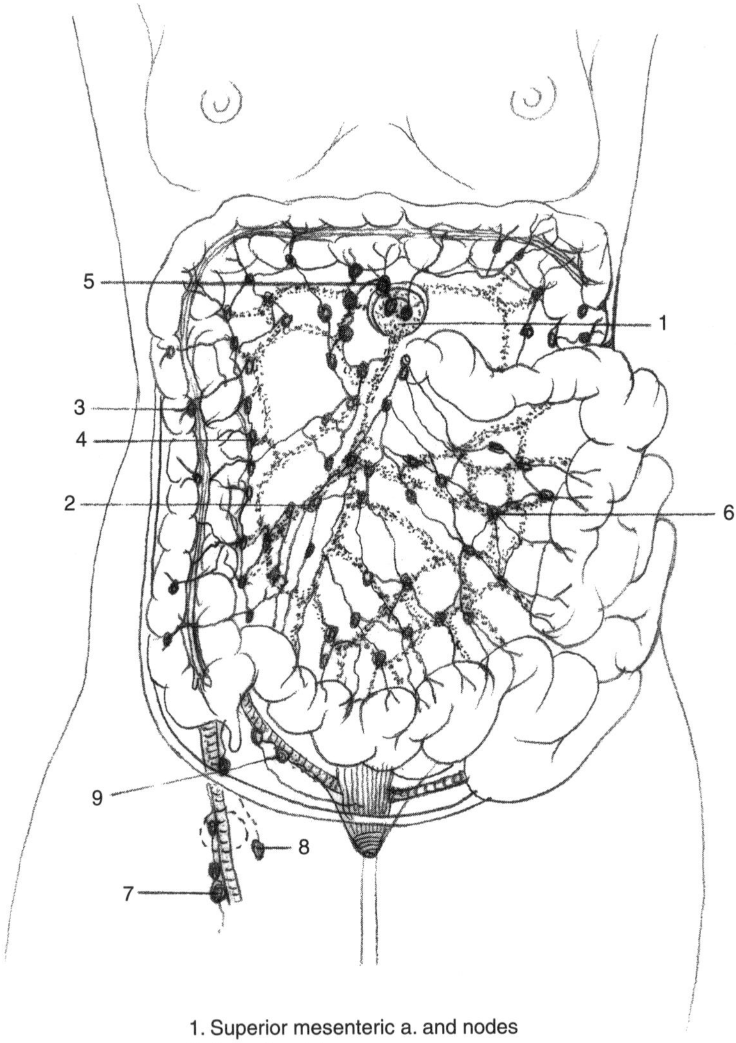

1. Superior mesenteric a. and nodes
2. Ileocolic node
3. Epicolic node
4. Paracolic node
5. Mesocolic node
6. Mesenteric node
7. Inferior superficial inguinal node
8. Deep inguinal node
9. Internal iliac node

30.2 Lymphatic drainage of abdomen

Figure 30.2: Lymphatic drainage of abdomen

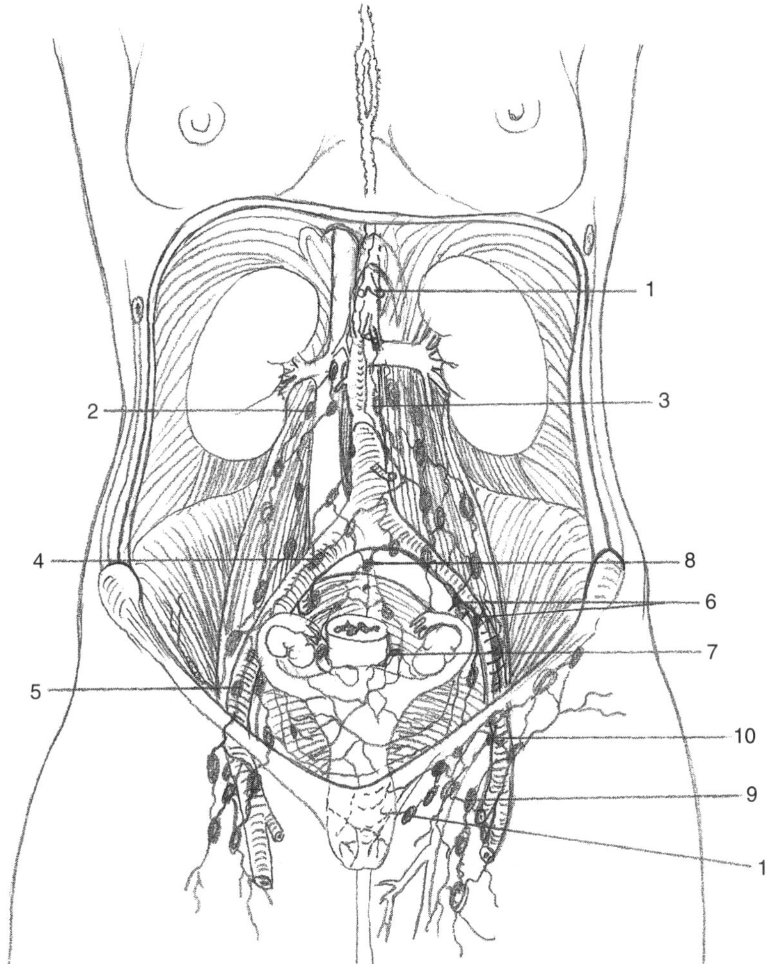

1. Thoracic duct
2. Lateral caval node
3. Lateral aortic node
4. Common iliac node
5. External iliac node
6. Internal iliac node
7. Pararectal node
8. Promontorial node
9. Inferior superficial inguinal node
10. Superolateral superficial inguinal node
11. Superomedial superficial inguinal node

30.3 Lymphatic drainage of abdominopelvic region

Figure 30.3: Lymphatic drainage of abdominopelvic region

Chapter 31
FUNCTIONAL ANATOMY OF THE HIP

First up, a discussion of the skeletal anatomy of the hip joint. Then, a description and examination of muscles acting on the joint. And, later, an evaluation of factors and forces, including gravity, influencing function at the hip, knee, and ankle during posture and locomotion.

The Skeletal Anatomy of the Hip Joint

The **hip joint** is an *enarthrosis*, *i.e.*, a synovial joint of the ball and socket variety. Though more stable than the shoulder, the hip joint, as befits a weight-bearing joint, is much less mobile (see also chapters 2 and 18).

Stability at the hip is assured primarily by the reciprocal shape of articular contact areas and their fitting or congruence. The **head of the femur**, the ball of the enarthrosis, nestles deeply in the **acetabulum** (L., *acetabulum* = a vinegar cup) of the hipbone (Fig. 31.1). The acetabular labrum, a fibrocartilaginous lip, further increases the depth of the socket. The acetabular fossa, however, is incomplete below; the gap, the acetabular notch, is bridged by the transverse acetabular ligament. Branches of the obturator and medial circumflex femoral arteries pass through the acetabular notch to the **fovea** (L., *fovea* = a pit) on the femoral head, carried there by the **ligament of the head of the femur**.

The hip bone, *os coxae*, is the composite of three separate bones that fuse at the time of puberty. Until then, the triradiate cartilage, resembling a three-pronged star, unites them in the floor of the acetabular fossa.

The upper end of the femur is formed by the head, neck, and greater trochanter. The **neck** is narrower in the middle than at either end. Being some 5 cm (2 in.) long, the neck pushes the limb away from the trunk; it, thus, performs the same function as the clavicle at the shoulder. The angle between femoral neck and shaft is roughly 125°; it is slightly less in the female, because the female pelvis is wider.

The **greater trochanter**, just below the skin, is palpable. A horizontal line connecting the tips of the greater trochanters is approximately level with the center of the hip joints, the ischial spines, the tip of the coccyx, and the upper margin of the pubic symphysis.

The capsule of the hip joint is thick and strong. It is reinforced by three ligaments, one each from the ilium, pubis, and ischium. The **iliofemoral ligament**, also known as the Y-ligament because of its shape, is the strongest ligament in the body (Fig. 31.2). It passes in front of the hip joint and tightens when the limb is extended, or when poor posture forces the pelvis into a backward roll. Think of it as providing antigravity support.

The Muscles of the Hip Joint

Muscles that flex the thigh, *e.g.*, iliopsoas or rectus femoris, lie in front of the joint and those that extend it, including biceps femoris, semitendinosus, and semimembranosus—the **hamstrings**—are behind the joint. Muscles that adduct the limb are in medial compartment of the thigh and abductors are in its lateral compartment.

Rotators of the hip are short muscles whose very important function is to act as stabilizing forces or guardians of the hip joint in a manner not unlike that of the rotator cuff muscles at the shoulder. Lateral rotators pass across the back of the joint, *i.e.*, behind the vertical axis around which the femoral head rotates, and medial rotators cross in front of the joint and axis of rotation. Lateral rotation at the hip, known as "turn-out" to ballet dancers, gives strength to the limb, increases its capacity for leverage, and is necessary for maximum extension.

Lateral rotation is a forceful movement, more so than medial rotation. It will come as no surprise, therefore, that there are many more lateral rotator than medial rotator muscles. Gluteus maximus, obturator externus and internus, piriformis, and quadratus femoris all are lateral rotators of the hip, whereas the only authentic medial rotator muscles are the anterior fibers of gluteus medius and minimus and the tensor fasciae latae.

A few words now regarding the tensor fasciae latae, a muscle whose functional significance is frequently overlooked.

The tensor originates from the tuberculum of the iliac crest and the anterior border of the ilium, and it inserts into the front of the iliotibial tract—the vertically oriented aponeurotic fibers lateral in the fascia lata of the thigh. Passing in front of the hip joint and lateral to it, the tensor fasciae latae is able to flex, abduct, and medially rotate the thigh. Its more important function, however, is to help maintain erect posture by counteracting the extensory force of gluteus maximus fibers that insert into the iliotibial tract from behind and exercise backward traction. If unopposed, gluteus maximus, acting through the iliotibial tract, rotates the pelvis backward, thereby causing the body to lose balance.

The tensor fasciae latae also provides antigravity support at the knee, for the iliotibial tract attaches below to the lateral condyle of the tibia and blends with the capsule of the knee joint. Acting from above, and aided by gluteus maximus, the muscle steadies the femoral condyle on the tibia and assists in keeping the knee joint extended.

The Hip Joint in Posture and Locomotion

In individuals assuming the ideal anatomical position, the line of gravity passes through the center of an axis connecting the two hip joints. And, for as long as ideal posture is maintained, no muscular energy is required to stabilize the joint, for the force of gravity is matched by an equal and opposite force acting from the ground up. Now consider a soldier standing stiffly at attention, his body leaning forward; or, think of the relaxed individual reclining on his heels. In the first case, the line of gravity descends in front of the hip joint axis, requiring contraction of hip extensors to prevent the body from jack-knifing. In the second example, the line of gravity passes behind the joint axis, necessitating hip flexor activity to save the body from falling backward.

Actually, of course, the line of gravity changes constantly, as we hardly ever stand still. Yet, surprisingly, electromyographic studies show many antigravity muscles to be inactive even as we shift positions. Most muscles contract only when we are about to lose our balance; evidently, fibrous capsules and ligaments are strong enough to act as antigravity support forces during normal upright posture.

When both feet are on the ground, each hip joint is subjected to a compression force equal to one-half the body weight. Naturally, the body will tilt sideways whenever one leg is off the ground, unless the gravitational pull is resisted by an equal and opposite force acting on the hip joint of the supporting limb. It can be calculated that the joint then sustains a compression force of almost four times the body weight. Such a load predisposes towards fracture of the femoral neck, especially in older individuals in whom osteoporosis has reduced bone mass.

The muscles that oppose the effects of gravity when one foot is off the ground are the abductors of the hip joint *on the weight-bearing side* (Fig. 31.4)—in particular, gluteus medius and minimus. Both muscles arise above from the gluteal surface of the ilium and insert below on the greater trochanter. Acting from their trochanteric (femoral) attachment, gluteus medius and minimus steady the pelvis by keeping the weight-bearing hip joint in a position of relative abduction (Fig. 31.5). Failure to control pelvic sag towards the unsupported side (known clinically as a positive Trendelenburg sign) may indicate abductor weakness due, perhaps, to myopathic or neuropathic disease.

During the normal walking cycle, defined as the interval between two successive heel strikes of the same limb (Fig 31.6), the two legs alternate in their phases of support and float. While one leg is on the ground and sustains the individual's body weight, the other swings freely and carries with it the forward momentum of the body. The hip joints, therefore, also alternate in their roles as stabilizers of the pelvis; and, if the abductors on the side of the weight-bearing hip function properly to neutralize the effects of gravity, pelvic equilibrium will be maintained as we move from place to place.

Gluteus medius and minimus, medial rotators of the thigh, perform a second vital function during the walking cycle. Acting from their trochanteric attachments, they rotate the pelvis towards the weight-bearing femur as the other limb goes through the swing phase to advance the body one step. The pelvis, therefore, moves in the course of one walking cycle twice: once from a position of relative lateral rotation to one of relative medial rotation; and, a second time, in reverse order.

Effects of Gravity on Knee and Ankle Joints

The line of gravity in individuals with ideal posture passes *through* the axis connecting the centers of the hip joints, but descends *in front* of the axis linking the two knee joints.

The **knee joint**, formed by greatly expanded ends of femur and tibia, is a complex synovial joint of the hinge variety with a limited capacity for rotation. Though the femur seems to rest rather precariously on the tibia, the joint is surprisingly strong. Its stability is derived primarily from a fibrotendinous capsule, both extracapsular and intra-articular ligaments, and from aponeurotic expansions of muscles crossing the joint.

The knee can be flexed by muscles that cross the joint posteriorly, *e.g.*, hamstring muscles, and extended by **quadriceps**, a four-headed muscle that passes anterior to the joint.

Knee extension, of course, is essential to upright posture—yet, in a person standing erect, the quadriceps muscles and tendon are relaxed (as you may verify by moving the patella back and forth). The solution to this riddle is quite simple, of course: in most individuals standing tall, the line of gravity falls in *front* of the knee joints; therefore, gravity acts as the extensory force. Only when you bend your knees or lean backward, thereby shifting the line of gravity to the back of the knee, do quadriceps muscles contract to keep the body from falling. Strange as it may seem, even individuals without functional quadriceps can stand erect.

The **ankle** (talocrural) **joint** is represented by the talus held in a vise-like grip by the distal ends of tibia and fibula. It is a synovial hinge joint adapted for dorsal and plantar flexion of the foot. Muscles whose tendons cross in front of the ankle are dorsiflexors of the joint; their contraction forces the foot to point up. Plantarflexors cause the foot to point down; their tendons pass behind the ankle.

The line of gravity, as shown in Fig. 31.3, strikes the ground roughly midway between the heel and ball of the foot, hence passes in font of the ankle joint. Gravity, therefore, unless opposed by a countervailing force, will bend the ankle joint into a position of relative dorsiflexion and destabilize it. Normally, of course, the plantar flexors at the back of the leg, *e.g.*, gastrocnemius and soleus, resist this action of gravity and prevent the body from collapsing at the ankle.

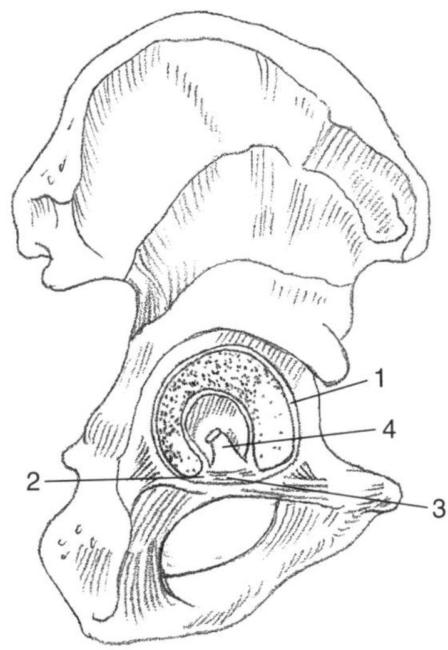

31.1 Acetabular fossa

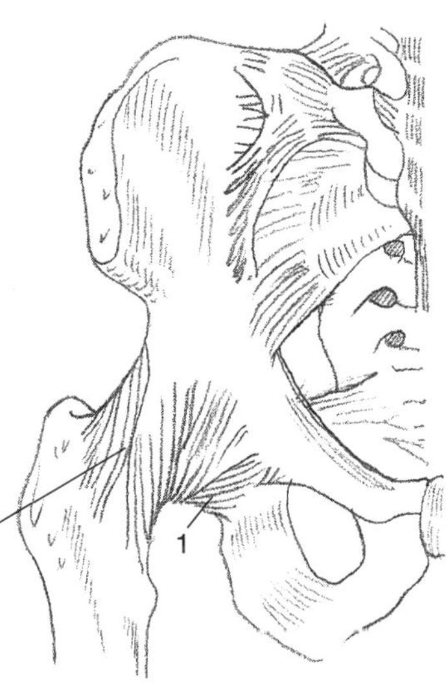

31.2 Support structures of hip joint

Figure 31.1: The acetabular fossa-skeletal anatomy of hip joint
Figure 31.2: Support structures of hip joint

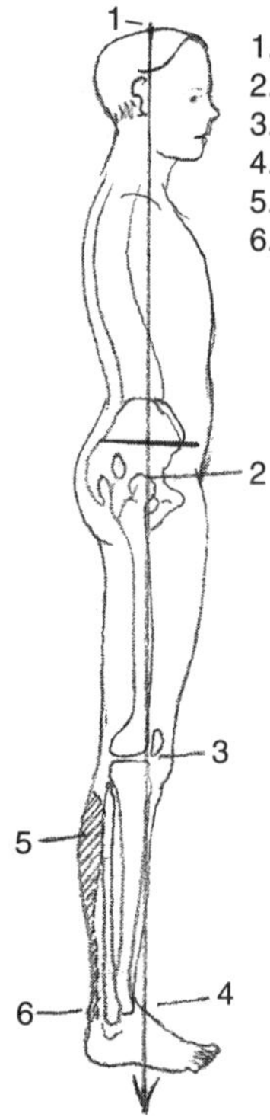

31.3 Line and force of gravity-erect posture

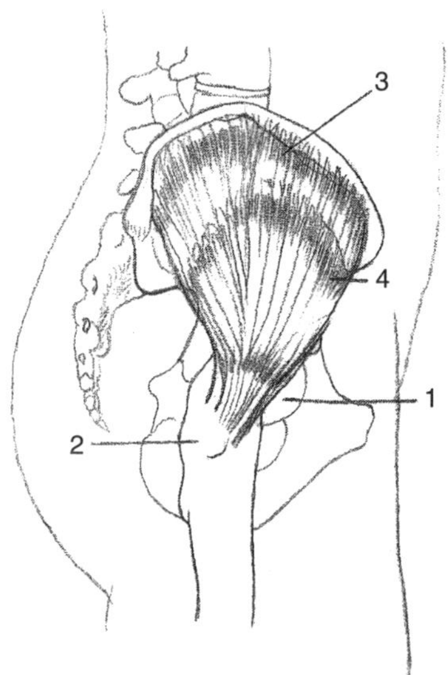

1. Head of femur in acetabular fossa
2. Greater trochanter
3. Gluteus medius m.
4. Gluteus minimus m.

31.4 Abductors and medial rotators at hip

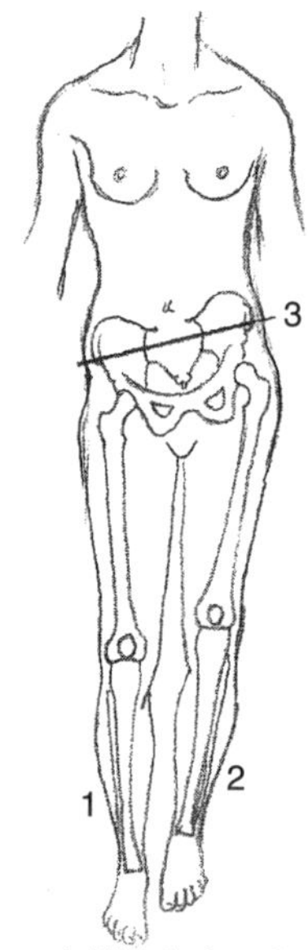

1. Weight-bearing limb
2. Floating limb
3. Anterior superior iliac spine

31.5 Action of hip abductors in normal gait

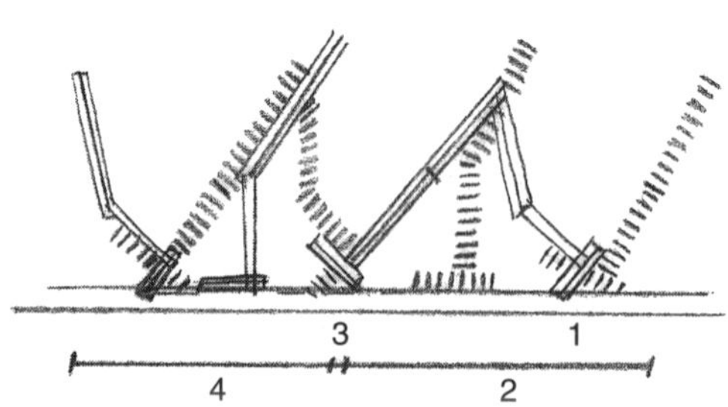

1. Heel strike
2. Stance phase
3. Toe-off
4. Swing phase

31.6 Walking cycle (right limb)

Figure 31.3: The line and force of gravity in erect posture
Figure 31.4: Abductors and medial rotators at the hip
Figure 31.5: Action of hip abductors in normal gait
Figure 31.6: The walking cycle

Chapter 32
INNERVATION OF LOWER LIMBS

As you will see, the innervation of the lower limb is perhaps slightly more complicated than that of the upper limb; but it, too, can be reduced to a set of principles. First, however, a review of the salient features of lower limb development to help explain the functional organization of the lumbosacral plexus.

The Fetal Lower Limb

As first discussed in chapter 20, the lower limb buds appear early in the fifth week of human development in the embryo's lumbosacral region. A septum divides each limb bud internally into anterior and posterior compartments. Muscles that develop in the anterior compartment, becoming flexors of the knee, ankle, and toes, are supplied by anterior divisions of the lumbosacral nerve plexus. The future extensors of the knee, ankle, and toes, forming in the posterior compartment, are innervated by posterior divisions of the limb plexus.

Limb compartments, surfaces, and borders become transposed after the fetal limb rotates medially about a vertical axis. Rotation explains why the pre-axial (tibial) border is on the medial side of the fully developed limb, and the postaxial (fibular) border on its lateral side. It also accounts for the fact that the flexor compartment faces posteriorly and the extensor compartment, anteriorly. For the same reason, anterior divisions of the lumbosacral plexus are behind the intermuscular septum and the posterior divisions of the plexus are in front of it.

The Lumbar Plexus

This nerve plexus (Fig. 32.1) is formed deep inside the psoas major muscle by anterior rami of the upper four lumbar nerves. Its branches are segmental body wall nerves, in series with the anterior rami of thoracic, *i.e.*, intercostal, nerves. They supply the skin and muscles of the lower abdominal wall, including iliopsoas and quadratus lumborum, the inguinal region, and the anterior thigh. They also carry sensory branches from the parietal peritoneum.

Iliohypogastric and ilio-inguinal nerves, branches of L1, run forward behind the anterior lamella of the thoracolumbar fascia. The **iliohypogastric nerve** becomes cutaneous a short distance above the superficial inguinal ring. The **ilio-inguinal nerve** runs below and parallel to the iliohypogastric nerve. Towards the end of its course through the lower abdominal wall, the nerve is inside the inguinal canal. Emerging from the superficial ring, its cutaneous branches innervate the root of the penis, the front and sides of the scrotum (mons pubis and labia majora), and the adjacent skin of the thigh.

The **genitofemoral** nerve, formed by anterior rami of L1 and L2, descends on the anterior surface of psoas major, deep to the psoas fascia. Just above the inguinal ligament, the nerve divides into genital and femoral branches. The **genital branch** enters the inguinal canal through the deep ring and supplies the cremaster muscle. The **femoral branch** accompanies the femoral artery into the thigh and, upon becoming cutaneous, innervates the skin below the intermediate one-third of the inguinal ligament.

The **lateral cutaneous nerve of the thigh**, derived from anterior rami L2 and L3, distributes to the skin over the vastus lateralis muscle.

The anterior rami of L4 and L5, along with the anterior rami of all sacral nerves, become the roots of the **lumbosacral plexus**. Roots split into anterior and posterior divisions, each dedicated to supply limb muscles according to their developmental origin anterior or posterior to the fetal intermuscular septum.

The *posterior divisions* of nerve roots L2–4 join to form the **femoral nerve**. Escaping from the lower, lateral border of psoas major, the femoral nerve runs deep to fascia iliaca, then breaks up into numerous branches, muscular and cutaneous, as it enters the front of the thigh. The nerve supplies all muscles that developed behind the fetal intermuscular septum. However, following medial rotation of the limb, the muscles come to lie anterior to it in the thigh (Fig. 32.2). They include not only the quadriceps muscles that extend the knee, but also sartorius which flexes it.

The *anterior divisions* of nerve roots L2–L4 unite to form the **obturator nerve**. It supplies a group of muscles that developed in front of the intermuscular septum. They, following internal rotation of the fetal limb, come to lie medially in the thigh and become its adductors. Other muscles forming in front of the fetal intermuscular septum, the future flexors of the knee, end up at the back of the thigh.

The anterior compartment is later subdivided by the appearance of a secondary septum, which separates future adductors and flexors. The obturator nerve and its branches supply all muscles in the adductor subcompartment of the thigh, and the sciatic nerve innervates the rest.

One final and important piece of information concerning the lumbar plexus: parts of L4 and all of L5 merge to form the **lumbosacral trunk**; it is the lumbar contribution to the lumbosacral plexus (Fig. 32.3).

The Lumbosacral Plexus

The lumbosacral plexus supplies all muscles and skin of the lower limb, gluteal region, and perineum. Its roots are the anterior rami of spinal nerves L4 through S3.

The **sciatic nerve** is the largest nerve of the lumbosacral plexus—indeed, it is the largest nerve in the body, containing close to 175,000 fibers. The nerve (L4–5, S1–3) enters the gluteal region through the greater sciatic foramen below the piriformis muscle (Fig. 32.4). It passes into the thigh between ischial tuberosity and greater trochanter and divides in the popliteal fossa into **tibial** and **common fibular** (peroneal) **nerves**.

Among the many collateral branches of the lumbosacral plexus are the superior and inferior gluteal nerves and the pudendal nerve. The **superior gluteal nerve** (L4–5, S1) supplies gluteus medius, gluteus minimus, and tensor fasciae latae—*i.e.*, muscles capable of abducting and medially rotating the thigh. The **inferior gluteal nerve** (L5, S1–2) innervates gluteus maximus, the largest muscle in the body. The **pudendal nerve** (S2–4) is the nerve of the perineum, distributing fibers to the anal triangle via its inferior anal branches, and to the urogenital triangle by way of the perineal branches, including the dorsal nerve of the penis or clitoris.

As mentioned earlier, each root of the lumbosacral plexus splits into anterior and posterior divisions (Fig. 32.3). Anterior and posterior divisions run side by side in the lumbosacral trunk and sciatic nerve, but separate as the nerve divides into tibial and common fibular branches. The tibial nerve carries with it the anterior divisions of roots L4 through S3, and the common fibular nerve takes away the posterior divisions of L4 through S2. Note that the most postaxial segment of the lumbosacral plexus, *i.e.*, S3, distributes exclusively to the flexor compartment of the limb. You will recall that T1, the lowermost segment of the brachial plexus, likewise, is represented only in the flexor compartment of the upper limb. And, like T1, which is the predominant source of nerve fibers to the intrinsic muscles of the hand, S3 fibers are committed to the muscles on the foot's plantar surface.

The Tibial Nerve and its Branches

The **tibial nerve**, the larger of the two terminal branches of the sciatic nerve, runs down the back of the leg and divides at the ankle into medial and lateral plantar nerves. It distributes to muscles at the back of the thigh and leg, and to the short muscles on the plantar aspect of the foot.

The branches of the tibial nerve innervate the hamstring muscles of the posterior thigh—semitendinosus, semimembranosus, and the long head, and *only the long head,* of biceps femoris—all extensors of the hip and flexors of the knee. Actually, because of their location proximal to the popliteal fossa, these muscles are not innervated by the tibial nerve directly, but by anterior (tibial) divisions of the sciatic nerve.

Below the knee, the tibial nerve supplies the superficial and deep muscles of the calf, such as gastrocnemius and soleus (triceps surae); also flexors digitorum and hallucis longus, the plantar flexors of the foot and long flexors of the toes.

The tibial nerve innervates all muscles on the plantar aspect of the foot through its terminal branches. The medial plantar nerve, the smaller of the two plantar nerves, innervates abductor hallucis (L., *hallux* = the great toe), the short flexors of all five digits, and the first lumbrical muscle. The lateral plantar nerve distributes all other intrinsic muscles on the foot's plantar surface.

The Common Fibular Nerve and Its Branches

The **common fibular** (peroneal) **nerve** passes behind the head of the fibula, loops around the neck of the bone, and splits within the fibularis (peroneus) longus muscle into superficial and deep fibular branches. Collectively, the fibular nerves distribute to muscles of the leg occupying the space in front of the interosseus membrane (Fig. 32.5). The anterior intermuscular septum divides that space into anterior and lateral compartments.

The deep fibular nerve supplies the muscles of the anterior leg compartment—tibialis anterior, extensors hallucis and digitorum longus, including fibularis tertius. Tibialis anterior is a dorsiflexor of the ankle joint; and, acting on transverse tarsal joints, an invertor of the foot—*i.e.*, it turns the sole of the foot inwards.

In addition, the deep fibular nerve innervates extensor digitorum brevis (and hallucis brevis), *the only muscle(s) on the dorsum of the foot.*

The superficial fibular nerve is the resident nerve of the lateral leg compartment. This compartment contains two muscles—fibularis longus and brevis; both are evertors of the foot, thus turn the sole of the foot outwards. But, because their tendons pass *behind* the lateral malleolus, the fibular muscles are also plantarflexors of the ankle. Cutaneous fibers of the superficial fibular nerve supply the skin of the lower leg in front, of the dorsum of the foot, and of most toes.

Once you have assimilated this fairly straight-forward scheme of nerve distribution, you need remember only the 1+1/2 muscle "rule" of the lower limb: One, the *short* head of biceps is the *only* muscle of the thigh innervated by common fibular (posterior) divisions of the sciatic nerve; and, two, extensor digitorum (and hallucis) brevis is the *only* muscle of the foot *not* innervated by the tibial nerve.

Summary of Nerve Supply to Lower Limb

Here, then, are the principal terminal branches of the lumbosacral plexus and their distribution:

1. **Femoral nerve** to the front of the thigh (flexors of the hip and extensors of the knee),
2. **Obturator nerve** to the medial (adductor) compartment of the thigh.
3. **Sciatic nerve** to all remaining muscles of the lower limb:
    a. **Tibial nerve** and its branches to the back of the thigh, the back of the leg (calf), and the plantar surface of foot
    b. **Common fibular nerve** and its branches to anterior and lateral leg compartments, *plus* 1+1/2 muscles; *i.e.,* the short head of biceps femoris in the thigh and the short digital extensor muscle on the dorsum of the foot

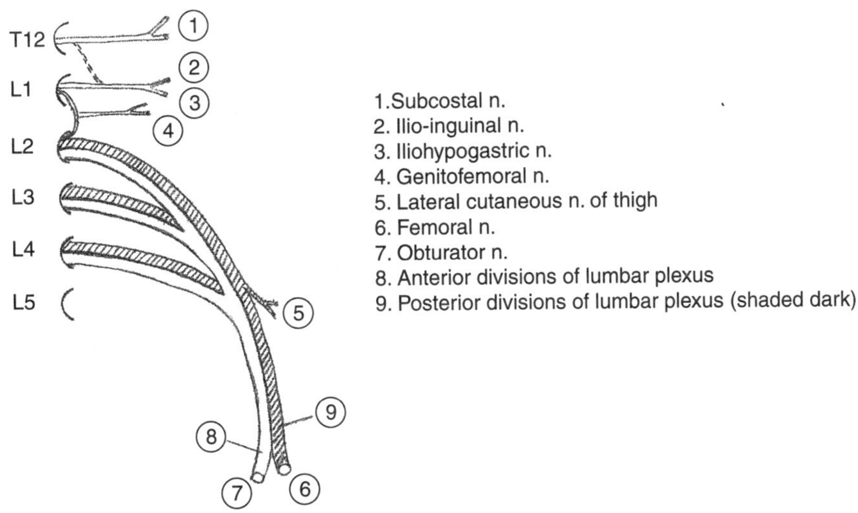

32.1 Anatomy of lumbar plexus

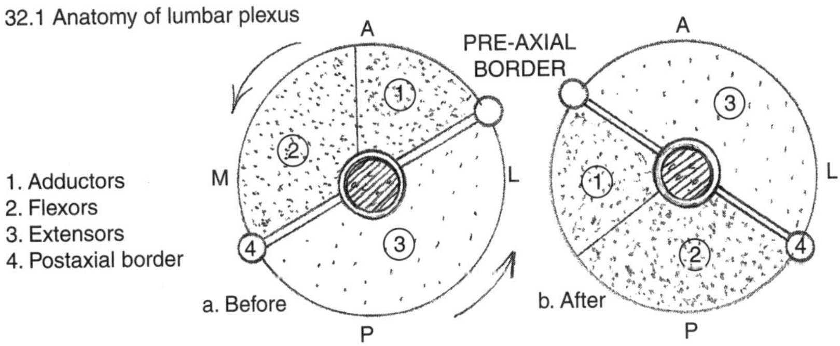

32.2 a-b Rotation of fetal lower limb (cross-section)

Figure 32.1: Anatomy of lumbar plexus
Figure 32.2: Schematic cross-section of fetal lower limb

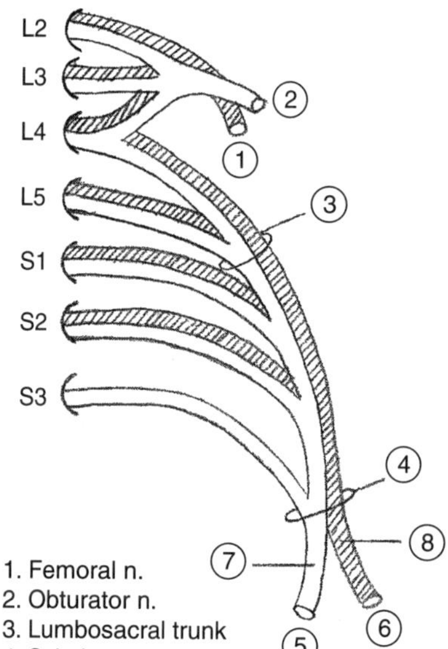

1. Femoral n.
2. Obturator n.
3. Lumbosacral trunk
4. Sciatic n.
5. Tibial n.
6. Common fibular n.
7. Anterior divisions of lumbosacral plexus
8. Posterior divisions of lumbosacral plexus (shaded)

32.3 Anatomy of lumbosacral plexus

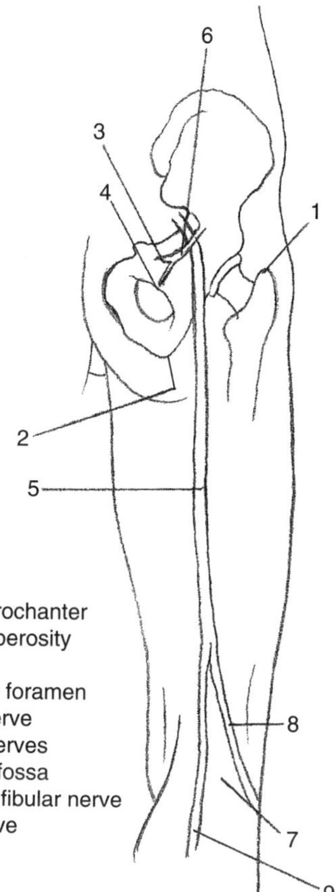

1. Greater trochanter
2. Ischial tuberosity
3. Pubis
4. Obturator foramen
5. Sciatic nerve
6. Gluteal nerves
7. Popliteal fossa
8. Common fibular nerve
9. Tibial nerve

32.4 Passage of sciatic nerve

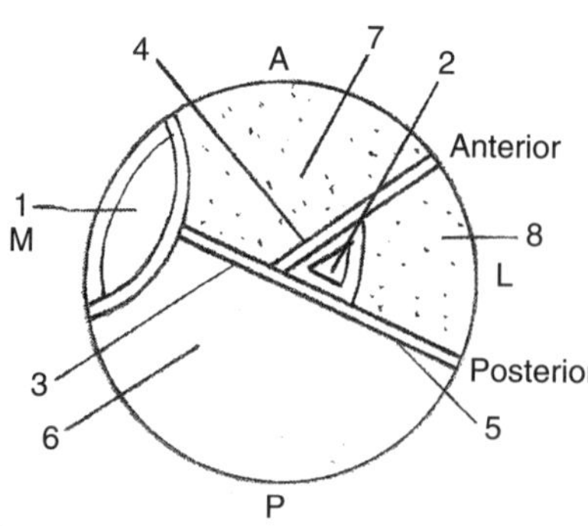

1. Tibia
2. Fibula
3. Interosseus membrane
4. Anterior intermuscular septum
5. Posterior intermuscular septum
6. Flexor compartment
7. Extensor compartment
8. Lateral compartment for fibularis mm.

32.5 Leg (cross-section)

Figure 32.3: Anatomy of the lumbosacral plexus
Figure 32.4: Passage and branches of the sciatic nerve
Figure 32.5: Schematic cross-section of the leg

INDEX